Intelligent Systems Reference Library

Volume 121

Series Editors

Janusz Kacprzyk, Polish Academy of Sciences, Warsaw, Poland

Lakhmi C. Jain, KES International, Shoreham-by-Sea, UK

The aim of this series is to publish a Reference Library, including novel advances and developments in all aspects of Intelligent Systems in an easily accessible and well structured form. The series includes reference works, handbooks, compendia, textbooks, well-structured monographs, dictionaries, and encyclopedias. It contains well integrated knowledge and current information in the field of Intelligent Systems. The series covers the theory, applications, and design methods of Intelligent Systems. Virtually all disciplines such as engineering, computer science, avionics, business, e-commerce, environment, healthcare, physics and life science are included. The list of topics spans all the areas of modern intelligent systems such as: Ambient intelligence, Computational intelligence, Social intelligence, Computational neuroscience, Artificial life, Virtual society, Cognitive systems, DNA and immunity-based systems, e-Learning and teaching, Human-centred computing and Machine ethics, Intelligent control, Intelligent data analysis, Knowledge-based paradigms, Knowledge management, Intelligent agents, Intelligent decision making, Intelligent network security, Interactive entertainment, Learning paradigms, Recommender systems, Robotics and Mechatronics including human-machine teaming, Self-organizing and adaptive systems, Soft computing including Neural systems, Fuzzy systems, Evolutionary computing and the Fusion of these paradigms, Perception and Vision, Web intelligence and Multimedia.

Indexed by SCOPUS, DBLP, zbMATH, SCImago.

All books published in the series are submitted for consideration in Web of Science.

More information about this series at https://link.springer.com/bookseries/8578

Gonçalo Marques · Alfonso González-Briones ·
José Manuel Molina López
Editors

Machine Learning for Smart Environments/Cities

An IoT Approach

Editors
Gonçalo Marques
Polytechnic of Coimbra
ESTGOH, Rua General Santos Costa
Oliveira do Hospital, Portugal

José Manuel Molina López
Universidad Carlos III DE Madrid
Colmenarejo, Spain

Alfonso González-Briones
Grupo de Investigación BISITE
Departamento de Informática y Automática
Facultad de Ciencias
Instituto de Investigación Biomédica de Salamanca
Universidad de Salamanca
Salamanca, Spain

ISSN 1868-4394 ISSN 1868-4408 (electronic)
Intelligent Systems Reference Library
ISBN 978-3-030-97518-0 ISBN 978-3-030-97516-6 (eBook)
https://doi.org/10.1007/978-3-030-97516-6

© The Editor(s) (if applicable) and The Author(s), under exclusive license to Springer Nature Switzerland AG 2022
This work is subject to copyright. All rights are solely and exclusively licensed by the Publisher, whether the whole or part of the material is concerned, specifically the rights of translation, reprinting, reuse of illustrations, recitation, broadcasting, reproduction on microfilms or in any other physical way, and transmission or information storage and retrieval, electronic adaptation, computer software, or by similar or dissimilar methodology now known or hereafter developed.
The use of general descriptive names, registered names, trademarks, service marks, etc. in this publication does not imply, even in the absence of a specific statement, that such names are exempt from the relevant protective laws and regulations and therefore free for general use.
The publisher, the authors and the editors are safe to assume that the advice and information in this book are believed to be true and accurate at the date of publication. Neither the publisher nor the authors or the editors give a warranty, expressed or implied, with respect to the material contained herein or for any errors or omissions that may have been made. The publisher remains neutral with regard to jurisdictional claims in published maps and institutional affiliations.

This Springer imprint is published by the registered company Springer Nature Switzerland AG
The registered company address is: Gewerbestrasse 11, 6330 Cham, Switzerland

Preface

The Internet of Things (IoT) consists of numerous "things" that are connected and managed across the Internet. Machine Learning is a research field that focuses on the development of predictive computer-aided applications which are not explicitly programmed. IoT and Machine Learning are mutually contributing to the advancement of technologies involved in smart city and smart environment development. Furthermore, the use of such communication and computing technologies is enabled by miniaturized microcontrollers, sensors, and actuators, which have a lower cost and greater energy efficiency.

Thus, IoT and ML provide smart cities and smart environments with countless advantages by enabling interaction with the physical world through interoperable applications, supported by smart city and smart environment systems. These applications aim to improve the citizens' quality of life.

This book aims to introduce Machine Learning and its applications for smart environments/cities supported by IoT technologies. Its chapters have been written by researchers located in different countries across the globe. The book consists of 12 chapters. Chapter 1 presents an introduction to the book scope and main topics. The rest of the chapters are organized into two parts. On the one hand, Part I comprises Chaps. 2–6 and focuses on smart environments. On the other hand, Part II comprises Chaps. 7–12 and presents relevant studies on smart cities. These chapters have been contributed by several authors from across the globe, namely Australia, Brazil, Colombia, Germany, Ghana, Italy, Malaysia, Portugal, Spain, Turkey, the Netherlands, and the USA. This makes the content of the book geographically diverse, as its authors come from 12 different countries, spread over 6 different continents.

The advent of the IoT and Machine Learning in recent decades has brought about developments in smart sensing and actuating technologies, which have been adopted in so-called smart environments, such as smart homes, smart farms, and other smart city settings. In this regard, keeping track of applications that use IoT and Machine Learning for smart environments has become an important aspect of research. A lot of research effort has been put into reviewing aspects of smart environments/cities, such as technologies, architectures, and security. However, there is not enough research on approaches that would combine IoT and Machine Learning in smart environments/cities. Chapter 1 "**An Introduction and Systematic Review on Machine Learning for Smart Environments/Cities: An IoT Approach**" presents a systematic review of the combination of IoT and Machine Learning in smart environments/cities. Furthermore, recommendations for the implementation of IoT and Machine Learning in smart environments/cities are presented. It is expected that the recommendations may be used as a basis for the successful implementation of IoT and Machine Learning in smart environments/cities.

Chapter 2 "**Model-Based Digital Threads for Socio-Technical Systems**" defines the MBSE-based methodology for the creation and maintenance of the digital thread of physical systems and their digital twins. The authors illustrate this with a case study in which the proposed methodology has been applied to design a digital thread of a Traffic Monitoring System (TMS) for a smart city. The methodology uses the SysML (Systems Modeling Language), which is adequate for the specification of socio-technological (cyber-physical) systems, such as TMSs. This digital thread represents both the physical and virtual entities of the system, enabling the development of digital twins for simulating, testing, monitoring, and/or maintaining the system. SysML is currently being redesigned, and the new SysML v2 aims to offer precise and expressive language capabilities to improve support to system specification, analysis, design, verification, and validation. The chapter also discusses how SysML v2 is expected to facilitate the development of digital threads for socio-technological systems, such as TMSs.

Since the beginning, smart cities have been predicated on the concept of IoT; however, there is a gap between theory and practice, making effective implementation challenging. Smart cities are not just for urban dwellers; they are also for suburban dwellers who need effective methods and systems to foster a higher quality of life. The principal pollutants that are the subject of this study are industrial suburbs. Chapter 3 "**IoT Regulated Water Quality Prediction Through Machine Learning for Smart Environments**" presents research on data from recycled wastewater obtained from industrial use, which would otherwise be dumped directly into rivers. The authors used IoT sensors to gather data which has enabled the inspection of water and the maintenance of its quality. The data were used to measure a series of factors that indicate and influence water quality; they were utilized to calculate the water quality index. Finally, three Machine Learning algorithms were used to train and predict this quality index.

Chapter 4 "**The Power of Augmented Reality for Smart Environments: An Explorative Analysis of the Business Process Management**" contains an explorative analysis of the use of augmented reality for smart environments. It demonstrates that augmented reality improves business process management in numerous fields, such as military, medicine, architecture, automotive, and retail. Additionally, the study furnishes historical evidence for the evolution of augmented reality for smart environments, reviews the contributions of top research in this field, and evidences the effect of augmented reality on the business process and on the transformation of the business environment into a smart environment through IoT devices. This type of devices is capable of monitoring important variables in the users' environments, from a value co-creation perspective. Indeed, there is evidence of the use of augmented reality in Information Management, Planning and Control, Change Process, Knowledge Management, Performance Management, People, and Customer Management. Finally, the chapter reveals how this technology can support innovative practices in business processes. In the search for competitive advantage, firm managers could indeed exploit this research to explore the impact of smart components on the improvement of smart environments investigating new industries.

IoT technologies are an opportunity for humanity to provide a wide range of e-educational technologies that have the potential to change educational systems. For example, students can now access online labs and libraries on their smart devices, which increases productivity in learning activities. Thanks to the rapid growth of information and communication technologies, online learning is now a practical choice, provided there is Internet availability. By combining real and virtual aspects in the learning process, IoT will allow for the expansion of learning contexts. The adoption of IoT can provide a wide range of e-educational technologies that have the potential to cover the shortfalls of the current education system. Through IoT, students can interact with their colleagues and teachers, share ideas, and find answers to problems. Chapter 5 "**Internet of Things Applications for Smart Environments**" considers IoT in education, the benefits of IoT technology in e-learning, online self-study, smart collaboration, IoT and e-learning, smart homes, smart home service adoption, and critical factors for smart home service.

A smart environment should be self-aware, using IoT technologies applied to health, education, and justice. Following the COVID-19 pandemic, the gains from applying technologies in healthcare and medical decision-making environments became evident. Chapter 6 "**Exploring Interpretable Machine Learning Methods and Biomarkers to Classifying Occupational Stress of the Health Workers**" focuses on the monitoring of stress in healthcare professionals through wearable devices, using biomarkers and Machine Learning to develop models that can aid in decision making. Challenges related to Explainable Artificial Intelligence are also addressed, as well as to the definition of stress classification, enabling the identification of impact on health professionals. An intelligent system is proposed to recommend actions in response to the professionals' stress level, in a way that is explainable, transparent, and feasible. This is an outstanding solution that may be adopted by the managers of health centers. Challenges related to information security and to the privacy of the health professionals are also discussed.

Chapter 7, "**Smart Cities, The Internet of Things, and Corporate Social Responsibility**" explores a plethora of IoT studies to identify how such technologies improve operational efficiency and infrastructure service and create an ecosystem in which the economic, environmental, and societal challenges associated with increased urbanization and smart cities may be addressed. Their inherent risks, issues, and challenges are also explored. Building on CSR literature, the author argues for a re-orientation of the smart city design toward decisional and governance process(es), and a shift away from technocentric and top-down approaches. A call is made for increased collaboration between decision-makers, community, and citizens in IoT implementation. A top-down/bottom-up multi-staged collaborative approach is proposed for evolving Corporate Social Responsibility governance and engagement. It recognizes the importance of creating shared value, in the selection and deployment of IoT devices. Consequentially, addressing and resolving the challenges faced by communities and citizens in the adoption of IoT in smart cities.

Chapter 8 "**Intelligent Techniques for Optimization, Modelling and Control of Power Management Systems Efficiency**" analyzes the issue of climate change and smart grid, from the point of view of efficiency in power converters. Power distribution in the Spanish electric system has changed over the years and will further change with the introduction of the smart grid, opening up new possibilities for the distribution of energy. Power electronics aim to interconnect the different parts of the electric system and control the energy flow from point to point. In addition, the main objective of smart cities is efficiency and the optimal use of energy. Power converters are a crucial element, as they interconnect electricity generators with electricity consumers. The introduction of Machine Learning and AI in this field helps to optimize the switching converter, allowing for the reduction of power loss.

Chapter 9 "**Intelligent Simulation and Emulation Platform for Energy Management in Buildings and Microgrids**" presents a Multi-Agent-based Real-Time Infrastructure for Energy (MARTINE). MARTINE is a platform that enables the study of the physical components of buildings and microgrids, including emulation capabilities, multi-agent and real-time simulation, and intelligent decision support models and services, based on Machine Learning approaches. MARTINE enables the study and management of energy resources, considering both physical and intelligent virtual components. Hence, it provides a real platform for the continuous improvement of the synergies between IoT and Machine Learning solutions.

Smart cities collect data using IoT technology and use the information obtained from this data to manage resources and services efficiently. Thus, the living conditions of the people living in the cities are facilitated by the quality services offered to them. Chapter 10 "**Machine Learning Applications and Security Analysis in Smart Cities**" presents smart city IoT applications; especially, smart parks, smart buildings, smart homes, smart health, smart business, and smart environment applications are widely used. It is possible to benefit from Machine Learning methods depending on the data obtained while developing these applications. With these methods, routine operations, especially for the city administration, can be made more practical and rational. However, while everything is smart, environmental influences should be taken into account. Systems that can be implemented in smart cities with the smart environment should be completely sensitive to the environment. The protocols used for the communication of data of IoT applications developed in smart environments and cities must be secure in terms of security.

Chapter 11 "**Recent Developments of Deep Learning in Future Smart Cities: A Review**" reviews the use of Deep Learning techniques in Artificial Intelligence applications, oriented toward multiple smart city domains, including smart transportation, smart services, smart governance, environment, security, and public safety. The difficulties associated using Deep Learning on smart city data have also been addressed.

This chapter is concluded by describing a series of Deep Learning techniques that help to better understand the smart city concept and follow the current trends in smart cities.

Chapter 12 "**Smart and Sustainable Cities in Collaboration with *IoT*: The Singapore Success Case**" explores the collaboration of the IoT paradigm with the Sustainable and Smart Cities (SSC) concept and looks at their success cases. The need for an adequate flow of information is emphasized, so that the state of a particular urban area may be known in real time. Accordingly, the challenges in the interconnection of highly sensitive sensors, as well as the transfer of data, require a hybrid cloud architecture that would allow for the large-scale processing of daily citizen data and for the prediction of environmental factors. However, the conceptualization and creation of an SSC must be considered in technological, scientific, social, and state policies, aspects that translate into Governance, Mobility, Sustainability, Economic Development, Intellectual Capital, and Quality of Life. Moreover, adding to the technological utopia, the modern concept of economic and social development entails the creation of an SSC for the promotion of entrepreneurship, innovation, and social justice: a new dimension of urban resilience focused on a city caring society.

Recent advances and comprehensive reviews have been included, aiming to provide background for future research initiatives. It is hoped that the book will be of support in the research and development of future IoT architectures for smart environments/cities. Finally, we thank everyone involved in this project for their contribution and for giving us the opportunity to edit this book. Furthermore, we would like to thank all the professionals from Springer who have worked with us, in particular, Prof. Lakhmi C. Jain, for their help and support during the development of this book.

Oliveira do Hospital, Portugal — Gonçalo Marques
Salamanca, Spain — Alfonso González-Briones
Colmenarejo, Spain — José Manuel Molina López

About This Book

This book introduces Machine Learning and its applications in smart environments/cities. At this stage, a comprehensive understanding of smart environment/city applications is critical for supporting future research. This volume includes chapters written by researchers from different countries across the globe and identifies critical threads in research and also gaps that open up new and challenging lines of research for the future. Recent advances are discussed, and thorough reviews introduce readers to critical domains. The discussion on key research topics presented in this book will accelerate smart city and smart environment implementations based on IoT technologies. Consequently, this book will support future research activities aimed at developing future IoT architectures for smart environments/cities.

Contents

About the Editors

Gonçalo Marques holds a Ph.D. in Computer Science Engineering and is Member of the Portuguese Engineering Association (Ordem dos Engenheiros). He is currently working as Assistant Professor lecturing courses on programming, multimedia, and database systems. Furthermore, he worked as Software Engineer in the Innovation and Development Unit of Groupe PSA automotive industry from 2016 to 2017 and in the IBM group from 2018 to 2019. His current research interests include Internet of Things, Enhanced Living Environments, Machine Learning, e-health, telemedicine, medical and healthcare systems, indoor air quality monitoring and assessment, and wireless sensor networks. He has more than 80 publications in international journals and conferences, is a frequent reviewer of journals and international conferences, and is also involved in several edited books projects.

Alfonso González-Briones holds a Ph.D. in Computer Engineering from the University of Salamanca since 2018; his thesis obtained the second place in the 1st SENSORS+CIRTI Award for the best national thesis in Smart Cities (CAEPIA 2018). At the same university, he obtained his Bachelor of Technical Engineer in Computer Engineering (2012), Degree in Computer Engineering (2013), and Master in Intelligent Systems (2014). He was Project Manager of Industry 4.0 and IoT projects in the AIR Institute, Lecturer at the International University of La Rioja (UNIR), and also a "Juan De La Cierva" Postdoc at University Complutense of Madrid. Currently, he is Assistant

Professor at the University of Salamanca in the Department of Computer Science and Automatics. He has published more than 30 articles in journals, more than 60 articles in books and international congresses and has participated in 10 international research projects. He is also Member of the scientific committee of the *Advances in Distributed Computing and Artificial Intelligence Journal* (ADCAIJ) and *British Journal of Applied Science and Technology* (BJAST), and a reviewer of international journals (*Supercomputing Journal*, *Journal of King Saud University*, *Energies*, *Sensors*, *Electronics or Applied Sciences* among others). He has participated as Chair and Member of the technical committee of prestigious international congresses (AIPES, HAIS, FODERTICS, PAAMS, KDIR).

José Manuel Molina López is Full Professor at the Universidad Carlos III de Madrid. He joined the Computer Science Department of the Universidad Carlos III de Madrid in 1993. Currently, he coordinates the Applied Artificial Intelligence Group (GIAA). His current research focuses on the application of soft computing techniques (NN, Evolutionary Computation, Fuzzy Logic, and Multi-agent Systems) to radar data processing, air traffic management, e-commerce, and ambient intelligence. He has authored up to 100 journal papers and 200 conference papers. He received a degree in Telecommunications Engineering in 1993 and a Ph.D. degree in 1997 both from the Universidad Politécnica de Madrid.

Chapter 1
An Introduction and Systematic Review on Machine Learning for Smart Environments/Cities: An IoT Approach

José Joaquín Peralta Abadía and Kay Smarsly

Abstract Over the last centuries, human activities have had a significant impact on the environment, usually damaging and exploiting land, water bodies, and air all around us. With the advent of the Internet of Things (IoT) and machine learning (ML) in recent decades, developments in smart sensing and actuating technologies have been adopted for the environment. As such, interactions between humans and the environment have become more synergetic and efficient, creating so-called "smart environments". In recent years, smart environments, such as smart homes, smart farms and smart cities, have matured at an increasing rate. Therefore, keeping track of applications for smart environments has become an important aspect of research. Although several research efforts have targeted reviewing aspects of smart environments, such as technologies, architectures, and security, a gap is identified. Reviews focusing on approaches using a combination of IoT and ML in smart environments/cities are lacking. In this chapter, a systematic review of the combination of IoT and ML in smart environments is presented. Moreover, a summary of approaches to combine IoT and ML in smart environments is provided. The findings achieved in this chapter materialize into recommendations for the implementation of IoT and ML in smart environments. It is expected that the recommendations may be used as a basis for successful implementations of IoT and ML in smart environments.

Keywords Smart cities · Internet of Things (IoT) · Machine learning · IoT applications · Smart environments

1.1 Introduction

Human activities have had an impact on the environment over the last centuries, damaging and exploiting land, water bodies, and the air around us. Urban areas cover around 2% of the surface of the earth while accounting for approximately 75%

J. J. Peralta Abadía (✉) · K. Smarsly
Institute of Digital and Autonomous Construction, Hamburg University of Technology, Hamburg, Germany
e-mail: joaquin.peralta@tuhh.de

© The Author(s), under exclusive license to Springer Nature Switzerland AG 2022

G. Marques et al. (eds.), *Machine Learning for Smart Environments/Cities*,
Intelligent Systems Reference Library 121,
https://doi.org/10.1007/978-3-030-97516-6_1

of the world's natural resources [1]. However, recent developments in smart sensing and actuating technologies, such as Internet of Things (IoT) and machine learning (ML) algorithms, are key enablers of monitoring and control activities, sensing and acting upon the environment. Interactions between humans and the environment have become more synergetic and efficient, creating so-called "smart environments". Smart environments have the ability to obtain knowledge from the surroundings and to act and adapt according to the needs of the inhabitants, improving the experience of living beings in the smart environments [2]. As such, smart environments, such as smart homes, smart farms and smart cities, have been developing at increasing growth rates in recent years. Therefore, keeping track of applications for smart environments has become an important aspect of research.

Several surveys have covered the usage of IoT in smart environments, focusing on different aspects and fields. Security aspects of smart environments, such as intrusion detection systems, have been reviewed, providing insights into security vulnerabilities in IoT architectures for smart environments [3]. Information and communication technology (ICT) aspects of smart environments have also been thoroughly reviewed, covering networking technologies and standards [2], platforms and frameworks [4], communication technologies and architectures [5], and the integration of augmented reality with IoT [6]. Other reviews have focused on specific fields of smart environments, such as the adoption of smart devices in waste management in smart cities [7]. Nevertheless, ML algorithms, representing a driving technology of smart environments, have not been the focus of reviews in the field of IoT and smart environments. Thus, a review focusing on approaches that combine IoT and ML, representing innovative cornerstones of smart environments, is lacking. In this chapter, research related to the combination of IoT and ML in smart environments/cities is reviewed. Approaches used to combine both technologies are summarized and compared, revealing insights into aspects of the implementation and efficiency.

The remainder of the chapter is structured as follows. Section 1.2 presents concepts related to smart environments and machine learning, as a basis for the review process. Section 1.3 summarizes and compares the results of the review of approaches that combine IoT and ML in smart environments. Section 1.4 discusses the results of the review and the implications on future research. Finally, Sect. 1.5 provides conclusions and potential future work obtained from the results of this chapter.

1.2 Smart Environments, Internet of Things, and Machine Learning Concepts

This section introduces the basic concepts necessary to perform the systematic review of the combination of IoT and ML in smart environments. First, smart environment concepts are elaborated. Thereupon, concepts of IoT and ML are presented.

1.2.1 Smart Environments

Smart environments are the link between computers and everyday settings and tasks, powered by recent advances in ML, IoT, and pervasive computing [8]. The four main features of smart environments are (i) remote control of devices, (ii) device communication, (iii) information acquisition and dissemination from intelligent sensor networks, and (iv) enhanced services provided by intelligent devices. Remote control of devices is the most basic feature of smart environments, freeing humans from the necessity of physically interacting with devices. Subsequently, device communication allows machine-to-machine (M2M) communication, using standardized protocols and retrieval of information from Internet sources, thus building informed models of smart environments. Following, information acquisition and dissemination from intelligent sensor networks provide the ability to perform automated adjustments in smart environments, based on sensor readings and device communication. Finally, enhanced services provided by intelligent devices offer advanced capabilities, such as washing machines equipped with smart sensors that determine appropriate washing cycle times.

Smart environments can be subdivided according to the application domains. The main application domains include smart cities, smart homes, smart buildings, smart health, smart grids, smart transportation, and smart industries (also referred to as “smart factories”) [2]. The application domains are characterized by varying characteristics, such as personal or business use and single-user or multi-user oriented [5]. Therefore, the aforementioned application domains will be taken during the review process to categorize studies involving the combination of IoT and ML in smart environments.

1.2.2 Internet of Things

The IoT is a paradigm in which IoT devices, denoted also as “things”, are interconnected in a worldwide network. As such, IoT is powered by the use of heterogeneous IoT devices, enabling the design of applications that involve virtually both humans and IoT devices [9]. A scalable, layered IoT architecture is necessary to connect the physical and the digital world, as IoT applications may involve a large number of heterogeneous IoT devices [10]. Figure 1.1 shows an example of an IoT architecture, based on the architecture proposed by Guth et al. [11], with an additional vertical layer of security, contemplating the recent needs in data privacy and security. The application layer serves the visualization and analysis of data as well as the control of the IoT devices of the sensing layer. The IoT integration middleware layer comprises data storage, business logic, and the definition of networking and communication protocols, as well additional services, such as alerts and machine learning algorithms. The sensing layer includes IoT edge gateways, IoT devices, and sensors and actuators, which interact with the real world.

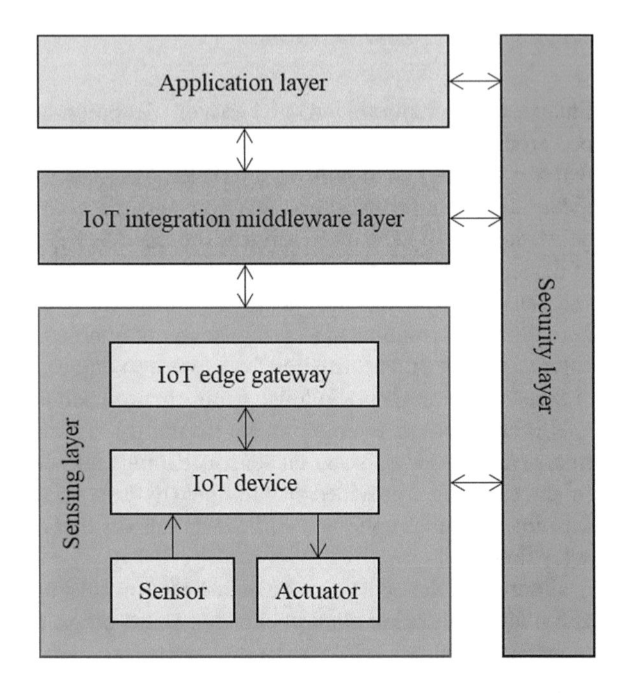

Fig. 1.1 IoT architecture with an additional security layer

1.2.3 Machine Learning

Machine learning is a subfield of computer sciences and a branch of artificial intelligence. Essentially, ML studies computer algorithms that learn and improve by using data, gaining experience, and generalizing knowledge [12]. ML algorithms observe data from datasets, build ML models based on the data and use the ML models to formulate hypotheses aiming to solve problems and to automate tasks [13]. Furthermore, ML algorithms, according to [13], may be categorized into three categories: (i) supervised learning, (ii) unsupervised learning, and (iii) reinforcement learning.

In *supervised learning* algorithms, datasets used to train ML models require a target feature, which can be labels in classification problems or numerical values in regression problems. In *unsupervised learning* algorithms, datasets used to train ML models lack a target feature and, as such, the ML model attempts to identify patterns and relationship between elements of the datasets (also referred to as datapoints). In *reinforcement learning* algorithms, outputs of the ML model are evaluated according to scoring functions and datasets are continuously updated with new datapoints, thus improving the knowledge of the algorithm through an iterative process. Finally, some ML algorithms combine supervised and unsupervised learning algorithms, called *semi-supervised learning* algorithms, and are capable of improving the performance of ML models in cases where only few datapoints are labeled. Consequently, in semi-supervised learning, only a small number of datapoints are labeled and a large number of datapoints are unlabeled.

Having presented the basic concepts of smart environments, IoT and ML, the following section presents the results of the review of the combination of IoT and ML in smart environments.

1.3 Results

This section reviews the combination of IoT and ML in smart environments. First, the methodology pursued during the review process is presented. Second, the quantity of studies reviewed herein, and the mean citation count are summarized for each smart environment domain. Finally, the review of the combination of IoT and ML in smart environments is detailed by domain, summarizing, and comparing the ML algorithms used in the studies, the IoT layer in which the ML algorithms are implemented, and the problem solved by the ML algorithms.

The review methodology follows three steps, (i) data collection, (ii) data organization, and (iii) data analysis. The data collection step involves searching for journal papers indexed in the Web of Science Core Collection as well as journal papers and conference papers indexed in the Scopus database. In an initial search, 1020 indexed studies, published between 2016 and 2021 and involving IoT, ML and smart environments, have been found, using the search string ((“smart environment*” OR “smart cit*” OR “smart home*” OR “smart grid*” OR “smart building*” OR “smart transportation” OR “smart health” OR “smart industr*” OR “smart factor*”) AND “machine learning” AND “IoT”). Next, the 100 most representative papers have been chosen, based on citation count. However, 29 papers lacked description of the ML algorithms used in the studies or presented the terms in the abstract but did not dwell in the topic in the full text. Consequently, the 29 studies have been omitted, entailing 71 studies to review. Then, in the data organization step, the IoT and ML characteristics described in the studies have been tabulated, according to the concepts presented in Sect. 1.2. Finally, an analysis of the organized data has been carried out, as presented in the remainder of this section.

Figure 1.2 presents the predominant terms in the abstracts of the studies in the form of a word cloud, where common multi-word terms in ICT have been joined as a single word. The primary predominant terms are “Internet of Things”, “data”, and “smart”, which highly relate to the topic being reviewed. Terms related to the field of machine learning follow in predominance, where “deep learning”, “machine learning”, “detection” and “model” have secondary predominance.

The papers reviewed in this study are grouped by domain, and the mean citation count is presented in Table 1.1, ordered by study count. It may be observed that smart cities and smart homes are the domains targeted most in smart environments, each with 18 studies, and the smart industries domain, with 2 studies, is the least targeted domain. Regarding the mean citation count, the most cited studies, with a mean of more than 40 citations, belong to the smart cities and smart grids domains, corroborating the interest of research and industry in both domains. Furthermore, smart transportation and smart homes follow, having a mean of more than 30 citations. The

Fig. 1.2 Predominant terms in the abstracts of the papers reviewed in this study

Table 1.1 Quantity of studies and mean citation count for each smart environment domain

Domain	Study count	Mean citations
Smart cities	18	44.6
Smart homes	18	31.6
Smart buildings	10	17.5
Smart health	9	26.1
Smart grids	7	40.4
Smart transportation	7	35.9
Smart industries	2	19.0

least cited studies belong to the smart buildings and smart industries domains. Having presented a general summary of the studies by domain, the following subsections present, by domain, a quantitative summary and comparison of the ML algorithms used in the studies, the IoT layer in which the ML algorithms are implemented, and the problem solved by the ML algorithms.

1.3.1 Smart Cities

Smart cities represent the broadest smart environment domain, encompassing the other smart environment domains. Therefore, studies involving the combination of IoT and ML in smart environments – with a broad covering scope, such as network intrusion detection and environmental monitoring – will mainly fall in this domain. Table 1.2 presents the studies targeting the smart cities domain grouped by machine learning type, data processing task, and algorithm. It may be noticed that supervised classification tasks have been preferred, where neural network algorithms and support vector machine algorithms are the most used ML algorithms, as the majority of

Table 1.2 Studies targeting the smart cities domain are grouped by machine learning type, data processing task, and algorithm

Learning type	Data processing task	Algorithm	Studies
Supervised	Classification	Neural networks	[14–19]
		Support vector machines	[14, 16, 20–25]
		Random forest	[20, 26]
		Decision tree	[16, 20]
	Regression	Neural networks	[27]
		Random forest	[27]
		Decision tree	[27, 28]
		Gradient boosting	[27]
		Logistic regression	[29]
		ARIMA	[30]
		Support vector regression	[31]
		Linear regression	[31]
Unsupervised	Clustering	DBSCAN	[23]
		k-means	[29]
	Feature selection	PCA	[17, 28]
		Auto-encoder	[16]

tasks involve labeling data, such as intrusion detection, object tracking, and user authentication.

The IoT layer in which the ML algorithms have been implemented in the studies targeting smart cities is summarized in Fig. 1.3.4, 6 It may be found that seven studies have implemented ML algorithms in the middleware layer. Furthermore, other seven studies have implemented ML algorithms in the sensing layer, where a preference for implementing ML algorithms in the IoT sensing edge is apparent. Regarding the problems to be solved by the ML algorithms, Fig. 1.4 reveals that the majority

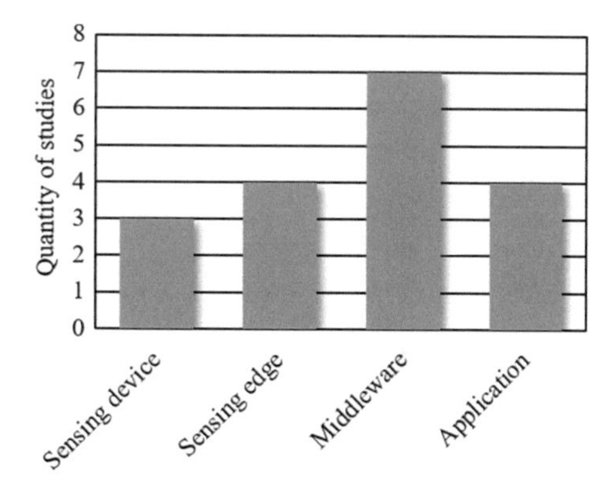

Fig. 1.3 IoT layer in which the ML algorithms are implemented

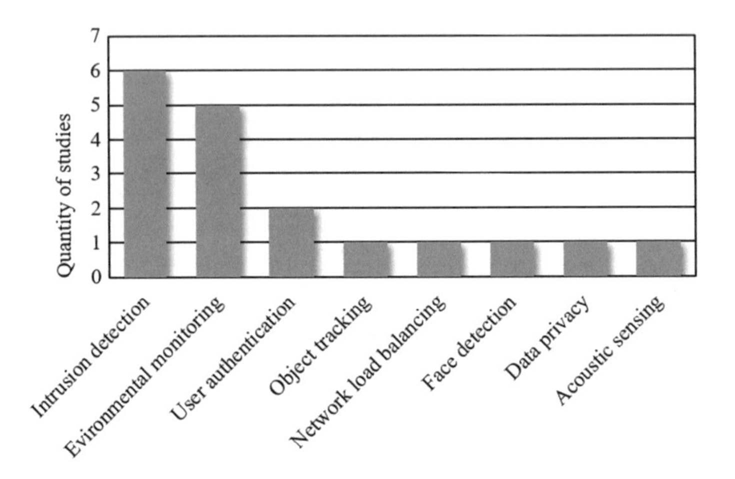

Fig. 1.4 Problems to be solved by the ML algorithms

of studies have targeted intrusion detection in the IoT networks and environmental monitoring, with six and five studies, respectively.

1.3.2 Smart Homes

Smart homes commonly refer to forms of residence, such as houses, apartments, or units in social housing developments. In the context of smart homes, IoT applications may be employed to sense and control devices for the wellbeing of the occupants, with applications including occupancy detection, object/person tracking, and device control [32]. Table 1.3 summarizes the studies targeting the smart homes domain

Table 1.3 Studies targeting the smart homes domain grouped by machine learning type, data processing task, and algorithm

Learning type	Data processing task	Algorithm	Studies
Supervised	Classification	Neural networks	[33–36]
		Support vector machines	[37]
		Random forest	[38–42]
		Decision trees	[42–44]
		Boltzmann machines	[45]
	Regression	Neural networks	[46, 47]
		Boltzmann machines	[45]
Unsupervised	Clustering	k-means	[40, 47, 48]
	Dimensionality reduction	Non-negative matrix factorization	[49]

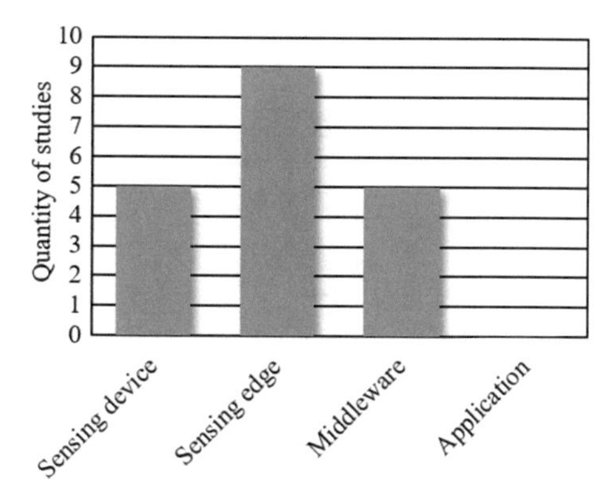

Fig. 1.5 IoT layer in which the ML algorithms are implemented

grouped by machine learning type, data processing task, and algorithm. It may be observed that supervised classification tasks have been preferred in smart home applications, where neural network algorithms and random forest algorithms have been the most used ML algorithms. A difference to the smart cities' domain may be noted, as random forest algorithms have been more popular in smart homes, probably due to the low memory and processing power needed by the algorithms.

The IoT layer in which the ML algorithms have been implemented in the studies targeting smart homes is summarized in Fig. 1.5. It may be found that five studies have implemented ML algorithms in the middleware layer, mainly in third-party cloud platforms. Furthermore, other 14 studies have implemented ML algorithms in the sensing layer, where a preference for implementing in the IoT sensing edge is once again apparent. As a special case, a study has devised an IoT application that uses ML in both the IoT edge and the IoT middleware for intrusion detection by means of anomalous activity detection [42]. Regarding the problems to be solved by the ML algorithms, Fig. 1.6 reveals that the majority of studies have targeted intrusion detection in the IoT networks and device identification, with four studies each. Following, activity recognition and energy management of devices have been targeted by the studies, with three studies each. In particular, a study has implemented an IoT application, which uses ML algorithms for device identification and energy management [45].

1.3.3 Smart Buildings

Smart buildings describe buildings that integrate materials, construction processes, intelligence, maintenance, and control as a single unit, based on the concept of adaptability and with the purpose of accomplishing energy management, efficiency, longevity, and comfort and satisfaction [50]. Table 1.4 presents the studies targeting the smart buildings domain grouped by machine learning type, data processing task, and algorithm. It may be observed that supervised regression tasks have been

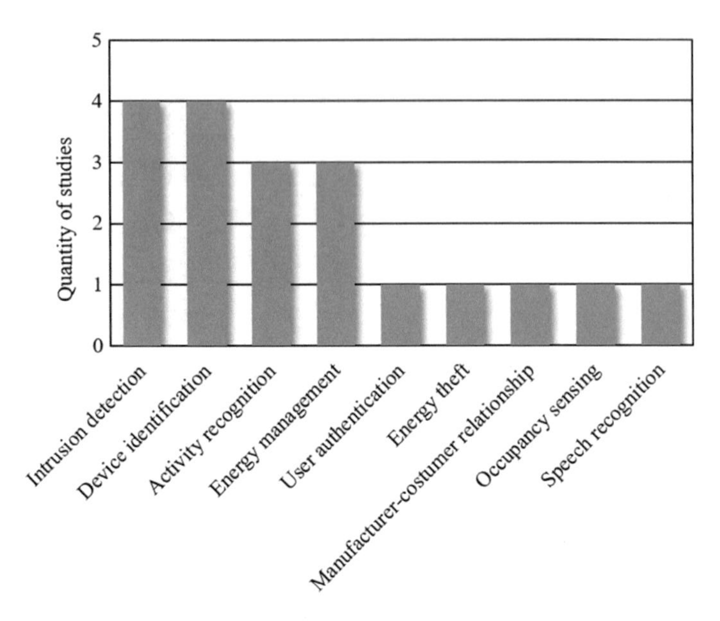

Fig. 1.6 Problems to be solved by the ML algorithms

Table 1.4 Studies targeting the smart buildings domain are grouped by machine learning type, data processing task, and algorithm

Learning type	Data processing task	Algorithm	Studies
Supervised	Classification	Neural networks	[51]
	Regression	Neural networks	[52–54]
		Random forest	[55]
		Support vector machines	[56]
		Logistic regression	[57]
		Decanter AI	[58]
		Artificial hydrocarbon networks	[59]
Unsupervised	Anomaly detection	Isolation forest	[60]
		Elliptic envelope	[60]
		Ensemble methods	[57]
Reinforcement			[52]

preferred in smart building applications, where neural network algorithms have been the most used ML algorithms.

The IoT layer in which the ML algorithms have been implemented in the studies targeting smart buildings is summarized in Fig. 1.7. It may be noticed that five studies have implemented ML algorithms in the application layer, mainly relying on historic data sensed for forecasting. Furthermore, four studies have implemented ML algorithms in the middleware layer and only one study has implemented an ML algorithm in the sensing layer. Regarding the problems to be solved by the ML algorithms, Fig. 1.8 reveals that the majority of studies have targeted energy management and thermal comfort, with three studies each, followed by occupancy detection tasks. From the ten studies, one study has implemented an IoT application,

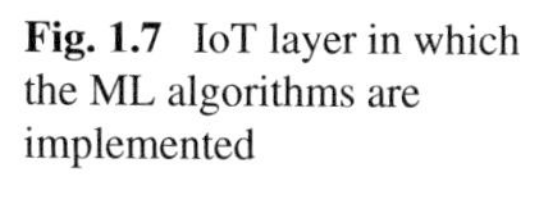

Fig. 1.7 IoT layer in which the ML algorithms are implemented

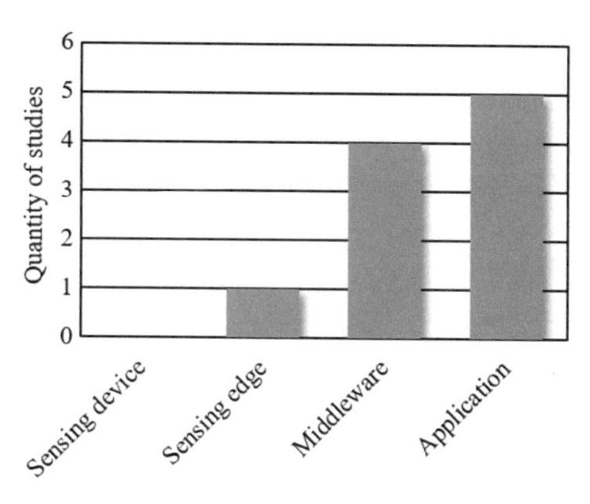

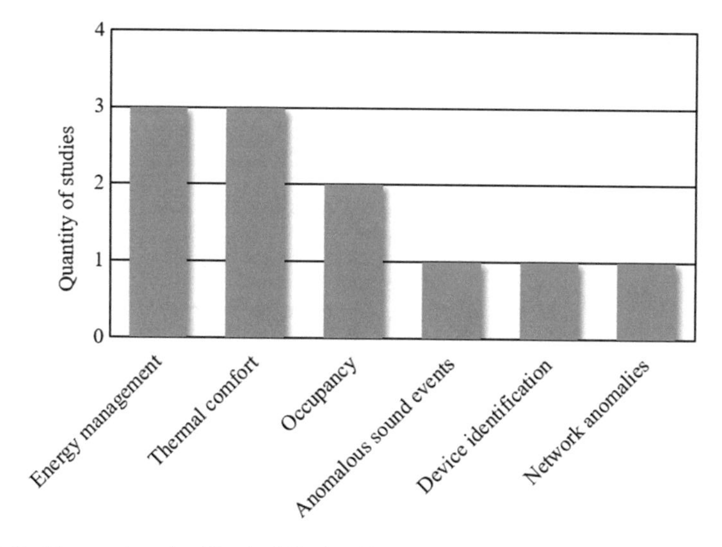

Fig. 1.8 Problems to be solved by the ML algorithms

which uses ML algorithms for device identification and network anomaly detection [57].

1.3.4 Smart Health

Smart health denotes the procurement of health services by means of context-aware networks and sensing technologies [61]. Table 1.5 presents the studies targeting the smart health domain grouped by machine learning type, data processing task, and algorithm. It may be found that supervised classification tasks have been preferred in smart health applications, where neural network algorithms and support vector machine algorithms have been the most used ML algorithms.

The IoT layer in which the ML algorithms have been implemented in the studies targeting smart health is summarized in Fig. 1.9. Most studies have implemented ML algorithms in the sensing layer, where a preference for implementing in the IoT sensing edge is once again apparent. Regarding the problems to be solved by the ML

Table 1.5 Studies targeting the smart health domain grouped by machine learning type, data processing task, and algorithm

Learning type	Data processing task	Algorithm	Studies
Supervised	Classification	Neural networks	[62–65]
		Support vector machines	[66–68]
		k nearest neighbors	[68]
		Decision tree	[69]
		Taylor Expanded Analog Forecasting Algorithm	[70]
	Regression	Taylor Expanded Analog Forecasting Algorithm	[70]

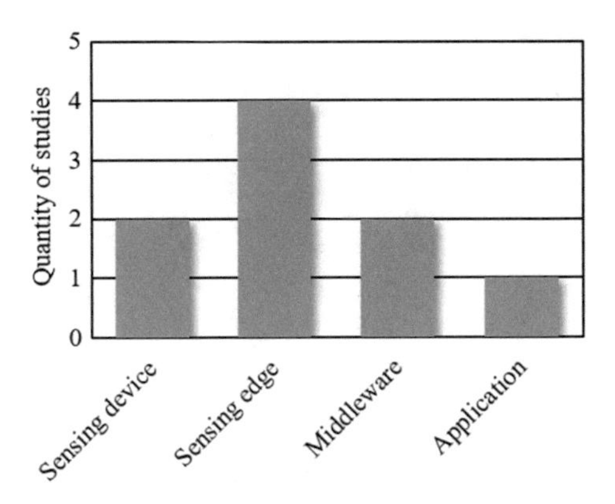

Fig. 1.9 IoT layer in which the ML algorithms are implemented

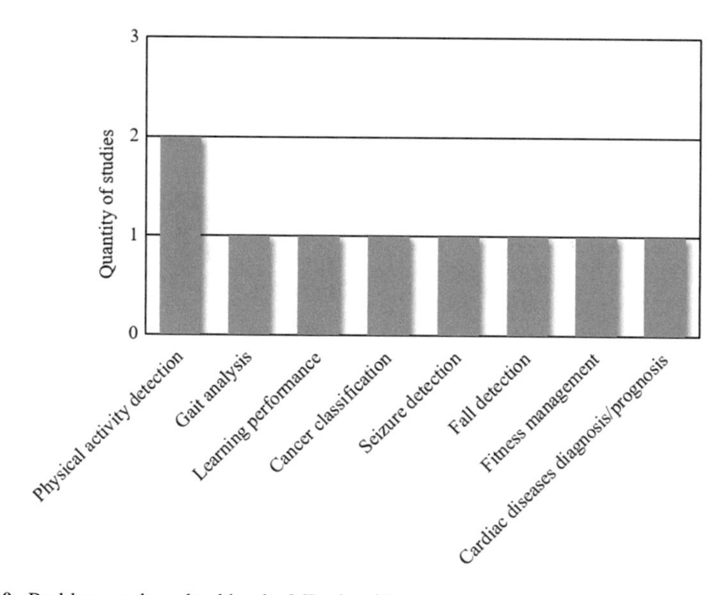

Fig. 1.10 Problems to be solved by the ML algorithms

algorithms, Fig. 1.8 reveals that the studies have targeted a varied array of problems, where no trend can be immediately detected (Fig. 1.10).

1.3.5 Smart Grids

Smart grids supply full visibility and pervasive control to energy providers over energy grids and energy services while empowering energy providers and customers with new ways of engaging with each other and performing energy transactions [71]. Table 1.6 presents the studies targeting the domain of the smart grid grouped by machine learning type, data processing task, and algorithm. It may be observed

Table 1.6 Studies targeting the domain of the smart grid grouped by machine learning type, data processing task, and algorithm

Learning type	Data processing task	Algorithm	Studies
Supervised	Classification	Naïve Bayes	[72, 73]
		Random forest	[73]
		Decision trees	[73, 74]
	Regression	Neural networks	[75–78]
		Decision trees	[74]
Unsupervised	Clustering	k-means	[77]
Reinforcement			[78]

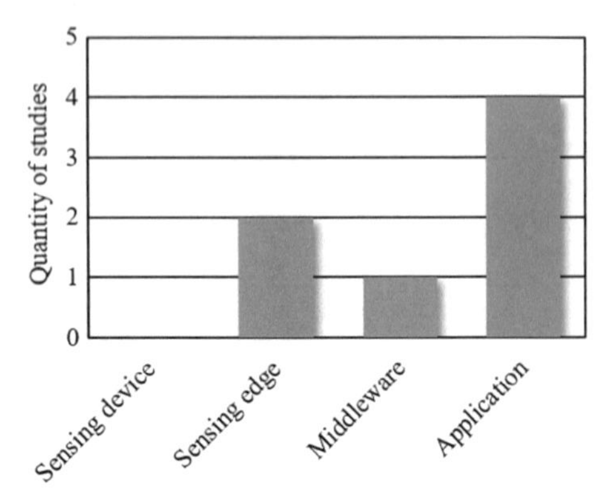

Fig. 1.11 IoT layer in which the ML algorithms are implemented

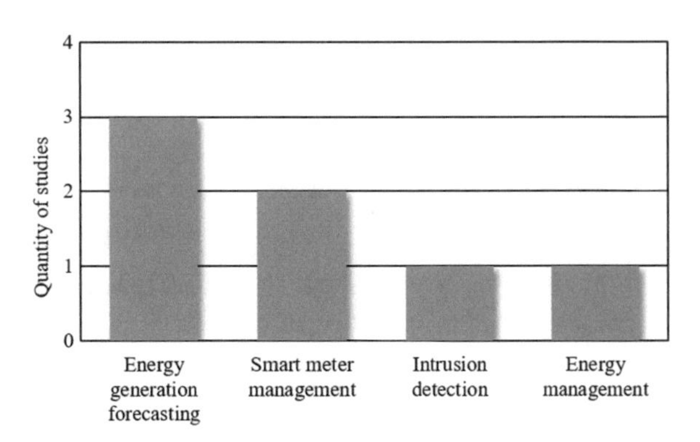

Fig. 1.12 Problems to be solved by the ML algorithm

that supervised regression tasks have been preferred in smart grid applications, where neural network algorithms have been the most used ML algorithms.

The IoT layer in which the ML algorithms have been implemented in the studies targeting smart grids is summarized in Fig. 1.11. Similar to smart buildings, most studies have implemented ML algorithms in the application layer, where forecasting tasks from historic data is performed. Regarding the problems to be solved by the ML algorithms, Fig. 1.12 reveals that the studies have targeted mainly energy forecasting and smart meter management.

1.3.6 Smart Transportation

Smart transportation aims to ensure road safety, efficient mobility, and reduced environmental impact, while updating users according to the location, based on instrumented and interconnected vehicular ad-hoc networks [79]. Table 1.7 presents the studies targeting the smart transportation domain grouped by machine learning type,

Table 1.7 Studies targeting the smart transportation domain grouped by machine learning type, data processing task, and algorithm

Learning type	Data processing task	Algorithm	Studies
Supervised	Classification	Neural networks	[80, 81]
		Support vector machines	[82]
		Random forest	[80, 83, 84]
		Decision trees	[80, 85]
		Naïve Bayes	[85]
		k nearest neighbors	[80, 85]
		Bayesian networks	[85]
		Ensemble methods	[80]
Unsupervised	Clustering	k-means	[86]

data processing task and algorithm. It may be noticed that supervised classification tasks have been preferred in smart transportation applications, without a clear preference for a specific algorithm.

The IoT layer in which the ML algorithms have been implemented in the studies targeting smart transportation is summarized in Fig. 1.13. Similar to smart buildings, most studies have implemented ML algorithms in the application layer, although smart transportation applications focus on classification tasks. Regarding the problems to be solved by the ML algorithms, Fig. 1.14 reveals that the studies have targeted mainly traffic management, being primarily performed in the application layer.

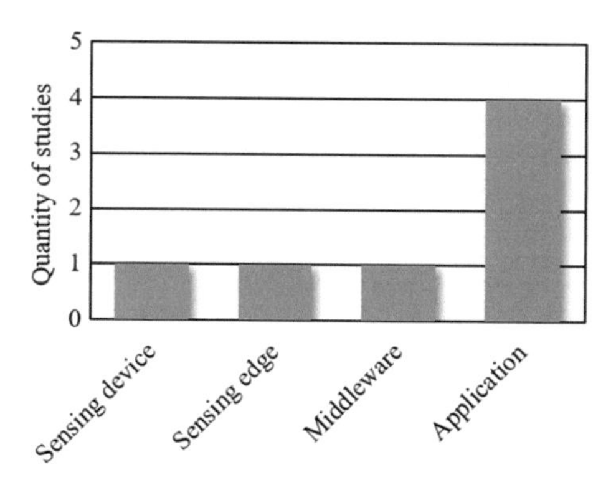

Fig. 1.13 IoT layer in which the ML algorithms are implemented

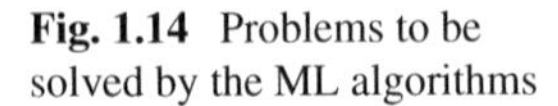

Fig. 1.14 Problems to be solved by the ML algorithms

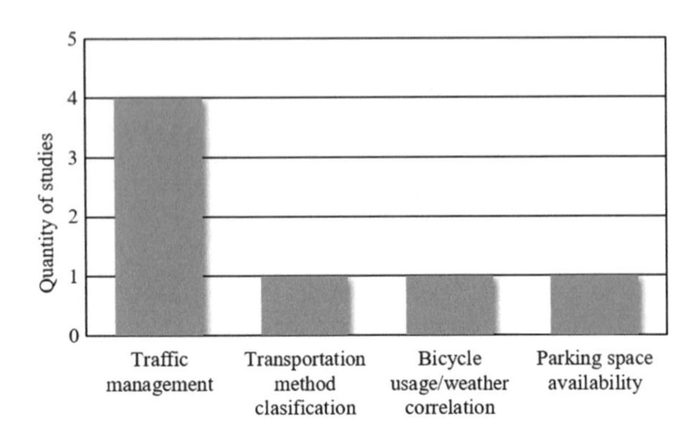

Table 1.8 Studies targeting the smart industries domain grouped by machine learning type, data processing task, and algorithm

Learning type	Data processing task	Algorithm	Studies
Supervised	Classification	Neural networks	[88]
		Decision trees	[89]
Unsupervised	Data generation	Auto-encoders	[88]

1.3.7 Smart Industries

Smart industries integrate ICT into the value chains, enabling better and personalized products and services by means of efficient, adaptive and flexible production, provisioning, and supply-chain processes [87]. Table 1.8 presents the studies targeting the smart industries domain grouped by machine learning type, data processing task and algorithm. It may be observed that supervised classification tasks, as well as unsupervised data generation tasks, have been preferred in smart industry applications.

The IoT layers in which the ML algorithms have been implemented in the studies targeting smart industries are the middleware layer and the application layer. In the middleware layer, automated defect inspection has been performed by means of neural networks and auto-encoders, while plant cultivation management has been performed in the application layer by means of decision trees. Having presented the review of the combination of IoT and ML in smart environments, the following section discusses the results of the review and the implications.

1.4 Discussion

This section presents the results of the survey of the combination of IoT and ML in smart environments. The findings and trends of the papers reviewed in this study are discussed, in an attempt to provide a strong foundation for successful implementations of ML algorithms in IoT applications in the context of smart environments.

Several trends have been noted as a result of the review presented in this study. First and foremost, there seems to be a preference in research advance benefits of citizens rather than of industry. Smart buildings and smart industries are the least cited domains, while smart cities and smart homes are the most published domains with a high mean citation count. Furthermore, research seems to have given much attention to the use of ML in IoT network security applications in the majority of smart environment domains, while there seems to be a lack of interest towards environmental monitoring and global warming, in spite of being a pressing matter in the last decades. In addition, it has been identified that the smart transportation domain and the smart industries domain are smart environment domains, which have lacked attention and require more research efforts to be further developed.

On the one hand, it has been found that studies targeting the smart homes and the smart health domains implement ML algorithms mainly in the IoT sensing layer, particularly in the sensing edge. Therefore, data privacy in the smart homes and the smart health domains has been able to be guaranteed, as sensitive data remains mostly in the sensing layer. In addition, the favored ML algorithms are lightweight in terms of processing power and memory usage, probably due to the usage of low-cost devices and low power availability. On the other hand, studies targeting the smart cities, the smart buildings, and the smart industries domains have a preference to implement ML algorithms in the IoT middleware layer, possibly due to high processing power availability of the cloud and ease of access to power sources and network resources. Also, the preferred ML algorithms in the smart cities, the smart buildings, and the smart industries domains are deep neural networks and support vector machines, which require more memory than other ML algorithms.

Regarding the preference of data processing tasks, it has been observed that the studies focus mainly on supervised learning algorithms. For supervised regression algorithms, studies have attempted to solve mostly forecasting problems, with a preference for implementing ML algorithms in the IoT application layer. For supervised classification algorithms, studies have focused mainly on labeling and identification problems, with a preference for implementing ML algorithms in the IoT sensing layer and IoT middleware layer. In addition, limited interest in unsupervised learning algorithms and reinforcement learning algorithms has been identified in the context of smart environments.

Finally, there seems to be a lack of research efforts towards decentralized ML algorithm implementations. Decentralized ML algorithms have been implemented to solve problems like intrusion detection [19] and activity recognition in IoT-wearable devices [36]. Nevertheless, much research effort is still needed to perform more efficient data transmission and processing of data in the IoT sensing layer. As such, decentralized technologies for ML algorithms, such as federated learning and blockchain-based crowdsourcing, are recommended to be adopted in smart environment applications, providing more efficient use of resources in resource-constrained devices.

1.5 Conclusions

Smart environments/cities provide knowledge from the surroundings, acting and adapting accordingly, i.e. "smart", to the needs of the inhabitants and improving the interactions between humans and the environment. Key elements in smart environments include IoT and ML algorithms, which provide the tools necessary for automated decision making. Despite its increasing importance in research and industry, reviews focusing on approaches using a combination of IoT and ML in smart environments have received little attention. Therefore, a systematic review of the combination of IoT and ML in smart environments has been presented in this chapter, summarizing and comparing the approaches used to combine both technologies.

As a result of this book chapter, small research interest in the smart transportation domain and smart industries domain has been noted. In addition, it has been identified that known problems, such as IoT network security, have been addressed thoroughly by research. Nevertheless, pressing problems of recent decades, i.e. environmental damage and global warming have not been addressed sufficiently. Furthermore, insights into aspects of the implementation and efficiency of the approaches have also been provided. Mainly, a lack of adoption of decentralized ML algorithms has been identified, with recommendations for future implementation of ML in IoT applications in the context of smart environments, such as the use of federated learning and blockchain-based crowdsourcing. Future research may be conducted to implement the recommendations proposed in this chapter, verifying and validating the suitability of the recommendations for the combination of IoT and ML and serving as a basis for successful implementations of IoT and ML in smart environments.

Acknowledgements This research is partially supported by the German Research Foundation (DFG) under grants SM 281/12-1 and SM 281/17-1. Any opinions, findings, conclusions or recommendations expressed in this chapter are those of the authors and do not necessarily reflect the view of DFG.

References

1. von Uexküll, J., Girardet, H.: Shaping our future: creating the world future council. Green books for the world future council initiative, Rev. and expanded edn. Totnes, United Kingdom (2005)
2. Ahmed, E., Yaqoob, I., Gani, A., Imran, M., Guizani, M.: Internet-of-things-based smart environments: state of the art, taxonomy, and open research challenges. IEEE Wirel. Commun. **23**, 10–16 (2016). https://doi.org/10.1109/MWC.2016.7721736
3. Elrawy, M.F., Awad, A.I., Hamed, H.F.A.: Intrusion detection systems for IoT-based smart environments: a survey. J. Cloud Comput. **7**, 21 (2018). https://doi.org/10.1186/s13677-018-0123-6
4. Alberti, A.M., Santos, M.A.S., Souza, R., Da Silva, H.D.L., Carneiro, J.R., Figueiredo, V.A.C., Rodrigues, J.J.P.C.: Platforms for smart environments and future internet design: a survey. IEEE Access **7**, 165748–165778 (2019). https://doi.org/10.1109/ACCESS.2019.2950656

5. Gomez, C., Chessa, S., Fleury, A., Roussos, G., Preuveneers, D.: Internet of Things for enabling smart environments: a technology-centric perspective. J. Ambient Intell. Smart Environ. **11**, 23–43 (2019). https://doi.org/10.3233/AIS-180509
6. Jo, D., Kim, G.J.: AR enabled IoT for a smart and interactive environment: a survey and future directions. Sens. (Basel) **19**, 4330 (2019). https://doi.org/10.3390/s19194330
7. Anagnostopoulos, T., Zaslavsky, A., Kolomvatsos, K., Medvedev, A., Amirian, P., Morley, J., Hadjieftymiades, S.: Challenges and opportunities of waste management in IoT-enabled smart cities: a survey. IEEE Trans. Sustain. Comput. **2**, 275–289 (2017). https://doi.org/10.1109/TSUSC.2017.2691049
8. Cook, D., Das, S.K.: Smart environments: technology, protocols, and applications. Wiley (2004)
9. Atzori, L., Iera, A., Morabito, G.: Understanding the Internet of Things: definition, potentials, and societal role of a fast evolving paradigm. Ad Hoc Netw. **56**, 122–140 (2017). https://doi.org/10.1016/j.adhoc.2016.12.004
10. Al-Fuqaha, A., Guizani, M., Mohammadi, M., Aledhari, M., Ayyash, M.: Internet of Things: A survey on enabling technologies, protocols, and applications. IEEE Commun. Surv. Tutor. **17**, 2347–2376 (2015). https://doi.org/10.1109/COMST.2015.2444095
11. Guth, J., Breitenbücher, U., Falkenthal, M., Leymann, F., Reinfurt, L.: Comparison of IoT platform architectures: a field study based on a reference architecture. In: Proceedings of the 2nd Cloudification of the Internet of Things, Paris, France, IEEE, 23 Oct 2016
12. Mitchell, T.M.: Machine learning, International edn, [Reprint.], McGraw-Hill, New York, NY, USA (2010)
13. Russell, S.J., Norvig, P.: Artificial intelligence: a modern approach, 4th edn. Pearson, Boston, MA, USA (2018)
14. Jain, R., Shah, H.: An anomaly detection in smart cities modeled as wireless sensor network. In: 2016 International Conference on Signal and Information Processing (IConSIP), Maharashtra State, India, IEEE, 10 June 2016, pp 1–5
15. Shafi, U., Mumtaz, R., Anwar, H., Qamar, A.M., Khurshid, H.: Surface water pollution detection using Internet of Things. In: 2018 15th International Conference on Smart Cities: Improving Quality of Life Using ICT & IoT (HONET-ICT), Islamabad, Pakistán, IEEE, 10/8/2018, pp 92–96
16. Rahman, M.A., Asyhari, A.T., Leong, L.S., Satrya, G.B., Hai Tao, M., Zolkipli, M.F.: Scalable machine learning-based intrusion detection system for IoT-enabled smart cities. Sustain. Cities Soc. **61,** 102324 (2020). https://doi.org/10.1016/j.scs.2020.102324
17. Bello, J.P., Mydlarz, C., Salamon, J.: Sound analysis in smart cities. In: Virtanen, T., Plumbley, M.D., Ellis, D. (eds.) Computational Analysis of Sound Scenes and Events, pp. 373–397. Springer International Publishing, Cham (2018)
18. Li, D., Deng, L., Lee, M., Wang, H.: IoT data feature extraction and intrusion detection system for smart cities based on deep migration learning. Int. J. Inf. Manage. **49**, 533–545 (2019). https://doi.org/10.1016/j.ijinfomgt.2019.04.006
19. Diro, A.A., Chilamkurti, N.: Distributed attack detection scheme using deep learning approach for Internet of Things. Futur. Gener. Comput. Syst. **82**, 761–768 (2018). https://doi.org/10.1016/j.future.2017.08.043
20. Naseer, M., Azam, M.A., Ul-Haq, M.E., Ejaz, W., Khalid, A.: ADLAuth: Passive authentication based on activity of daily living using heterogeneous sensing in smart cities. Sens. (Basel) **19**, 2466 (2019). https://doi.org/10.3390/s19112466
21. Zhang, H., Zhang, Z., Zhang, L., Yang, Y., Kang, Q., Sun, D.: Object Tracking for a smart city using IoT and edge computing. Sens. (Basel) **19**, 1987 (2019). https://doi.org/10.3390/s19091987
22. Sajjad, M., Nasir, M., Muhammad, K., Khan, S., Jan, Z., Sangaiah, A.K., Elhoseny, M., Baik, S.W.: Raspberry Pi assisted face recognition framework for enhanced law-enforcement services in smart cities. Futur. Gener. Comput. Syst. **108**, 995–1007 (2020). https://doi.org/10.1016/j.future.2017.11.013
23. Anjomshoa, F., Aloqaily, M., Kantarci, B., Erol-Kantarci, M., Schuckers, S.: Social behaviometrics for personalized devices in the Internet of Things era. IEEE Access **5**, 12199–12213 (2017). https://doi.org/10.1109/ACCESS.2017.2719706

24. Jan, S.U., Ahmed, S., Shakhov, V., Koo, I.: Toward a lightweight intrusion detection system for the Internet of Things. IEEE Access **7**, 42450–42471 (2019). https://doi.org/10.1109/ACCESS.2019.2907965
25. Shen, M., Tang, X., Zhu, L., Du, X., Guizani, M.: Privacy-preserving support vector machine training over blockchain-based encrypted IoT data in smart cities. IEEE Internet Things J. **6**, 7702–7712 (2019). https://doi.org/10.1109/JIOT.2019.2901840
26. Alrashdi, I., Alqazzaz, A., Aloufi, E., Alharthi, R., Zohdy, M., Ming, H.: AD-IoT: anomaly detection of IoT cyberattacks in smart city using machine learning. In: 2019 IEEE 9th Annual Computing and Communication Workshop and Conference (CCWC), Las Vegas, NV, USA, IEEE, 1/7/2019, pp. 305–310
27. Ameer, S., Shah, M.A., Khan, A., Song, H., Maple, C., Islam, S.U., Asghar, M.N.: Comparative analysis of machine learning techniques for predicting air quality in smart cities. IEEE Access **7**, 128325–128338 (2019). https://doi.org/10.1109/ACCESS.2019.2925082
28. Gomez, C.A., Shami, A., Wang, X.: Machine learning aided scheme for load balancing in dense IoT networks. Sens. (Basel) **18**, 3779 (2018). https://doi.org/10.3390/s18113779
29. Vu, D., Kaddoum, G.: A waste city management system for smart cities applications. In: 2017 Advances in Wireless and Optical Communications (RTUWO), Riga, Latvia, IEEE, 11/2/2017, pp 225–229
30. Thu, M.Y., Htun, W., Aung, Y.L., Shwe, P.E.E., Tun, N.M.: Smart air quality monitoring system with LoRaWAN. In: 2018 IEEE International Conference on Internet of Things and Intelligence System (IOTAIS), Bali, Indonesia, IEEE, 11/1/2018, pp 10–15
31. Roldán, J., Boubeta-Puig, J., Luis Martínez, J., Ortiz, G.: Integrating complex event processing and machine learning: an intelligent architecture for detecting IoT security attacks. Expert Syst. Appl. **149**, 113251 (2020). https://doi.org/10.1016/j.eswa.2020.113251
32. Balta-Ozkan, N., Davidson, R., Bicket, M., Whitmarsh, L.: Social barriers to the adoption of smart homes. Energy Policy **63**, 363–374 (2013). https://doi.org/10.1016/j.enpol.2013.08.043
33. Zhao, Y., Zhao, J., Jiang, L., Tan, R., Niyato, D., Li, Z., Lyu, L., Liu, Y.: Privacy-preserving blockchain-based federated learning for IoT devices. IEEE Internet Things J. **8**, 1817–1829 (2021). https://doi.org/10.1109/JIOT.2020.3017377
34. Roux, J., Alata, E., Auriol, G., Nicomette, V., Kaaniche, M.: Toward an intrusion detection approach for IoT based on radio communications profiling. In: 2017 13th European Dependable Computing Conference (EDCC), Geneva, Italy, IEEE, 9/4/2017, pp 147–150 (2017)
35. Chauhan, J., Seneviratne, S., Hu, Y., Misra, A., Seneviratne, A., Lee, Y.: Breathing-based authentication on resource-constrained IoT devices using recurrent neural networks. Computer **51**, 60–67 (2018). https://doi.org/10.1109/MC.2018.2381119
36. Bianchi, V., Bassoli, M., Lombardo, G., Fornacciari, P., Mordonini, M., de Munari, I.: IoT wearable sensor and deep learning: an integrated approach for personalized human activity recognition in a smart home environment. IEEE Internet Things J. **6**, 8553–8562 (2019). https://doi.org/10.1109/JIOT.2019.2920283
37. Ismail, A., Abdlerazek, S., El-Henawy, I.M.: Development of smart healthcare system based on speech recognition using support vector machine and dynamic time warping. Sustainability **12**, 2403 (2020). https://doi.org/10.3390/su12062403
38. Moriya, K., Nakagawa, E., Fujimoto, M., Suwa, H., Arakawa, Y., Kimura A., Miki, S., Yasumoto, K.: Daily living activity recognition with ECHONET Lite appliances and motion sensors. In: 2017 IEEE International Conference on Pervasive Computing and Communications Workshops (PerCom Workshops), Kona, HI, USA, IEEE, 3/13/2017, pp. 437–442 (2017)
39. Pinheiro, A.J., de Bezerra, J.M., Burgardt, C.A., Campelo, D.R.: Identifying IoT devices and events based on packet length from encrypted traffic. Comput. Commun. **144**, 8–17 (2019). https://doi.org/10.1016/j.comcom.2019.05.012
40. Thangavelu, V., Divakaran, D.M., Sairam, R., Bhunia, S.S., Gurusamy, M.: DEFT: a distributed IoT fingerprinting technique. IEEE Internet Things J. **6**, 940–952 (2019). https://doi.org/10.1109/JIOT.2018.2865604
41. Shahid, M.R., Blanc, G., Zhang, Z., Debar, H.: IoT Devices Recognition through network traffic analysis. In: 2018 IEEE International Conference on Big Data (Big Data), Seattle, WA, USA, IEEE, 12/10/2018, pp 5187–5192

42. Ullah, I., Mahmoud, Q.H.: A two-level flow-based anomalous activity detection system for IoT networks. Electronics **9**, 530 (2020). https://doi.org/10.3390/electronics9030530
43. Anthi, E., Williams, L., Slowinska, M., Theodorakopoulos, G., Burnap, P.: A supervised intrusion detection system for smart home IoT devices. IEEE Internet Things J. **6**, 9042–9053 (2019). https://doi.org/10.1109/JIOT.2019.2926365
44. Machorro-Cano, I., Alor-Hernández, G., Paredes-Valverde, M.A., Rodríguez-Mazahua, L., Sánchez-Cervantes, J.L., Olmedo-Aguirre, J.O.: HEMS-IoT: a big data and machine learning-based smart home system for energy saving. Energies **13**, 1097 (2020). https://doi.org/10.3390/en13051097
45. Mocanu, D.C., Mocanu, E., Nguyen, P.H., Gibescu, M., Liotta, A.: Big IoT data mining for real-time energy disaggregation in buildings. In: 2016 IEEE International Conference on Systems, Man, and Cybernetics (SMC), Budapest, Hungary, IEEE, 10/9/2016, pp. 3765–3769
46. Li, W., Logenthiran, T., Phan, V.-T., Woo, W.L.: A novel smart energy theft system (SETS) for IoT-based smart home. IEEE Internet Things J. **6**, 5531–5539 (2019). https://doi.org/10.1109/JIOT.2019.2903281
47. Li, W., Logenthiran, T., Phan, V.-T., Woo, W.L.: Implemented IoT-based self-learning home management system (SHMS) for Singapore. IEEE Internet Things J. **5**, 2212–2219 (2018). https://doi.org/10.1109/JIOT.2018.2828144
48. Grgurić, A., Mošmondor, M., Huljenić, D.: The smarthabits: an intelligent privacy-aware home care assistance system. Sens. (Basel) **19**, 907 (2019). https://doi.org/10.3390/s19040907
49. Yang, J., Zou, H., Jiang, H., Xie, L.: Device-free occupant activity sensing using WiFi-enabled IoT devices for smart homes. IEEE Internet Things J. **5**, 3991–4002 (2018). https://doi.org/10.1109/JIOT.2018.2849655
50. Buckman, A.H., Mayfield, M., Beck, S.B.M.: What is a smart building? Smart Sustain. Built Environ. **3**, 92–109 (2014). https://doi.org/10.1108/SASBE-01-2014-0003
51. Elsisi, M., Tran, M.-Q., Mahmoud, K., Lehtonen, M., Darwish, M.M.F.: Deep learning-based Industry 4.0 and Internet of Things towards effective energy management for smart buildings. Sens. (Basel) **21**, 1038 (2021). https://doi.org/10.3390/s21041038
52. Hu, W., Wen, Y., Guan, K., Jin, G., Tseng, K.J.: iTCM: toward learning-based thermal comfort modeling via pervasive sensing for smart buildings. IEEE Internet Things J. **5**, 4164–4177 (2018). https://doi.org/10.1109/JIOT.2018.2861831
53. Zhang, W., Hu, W., Wen, Y.: Thermal comfort modeling for smart buildings: a fine-grained deep learning approach. IEEE Internet Things J. **6**, 2540–2549 (2019). https://doi.org/10.1109/JIOT.2018.2871461
54. Chammas, M., Makhoul, A., Demerjian, J.: An efficient data model for energy prediction using wireless sensors. Comput. Electr. Eng. **76**, 249–257 (2019). https://doi.org/10.1016/j.compeleceng.2019.04.002
55. González-Vidal, A., Jiménez, F., Gómez-Skarmeta, A.F.: A methodology for energy multivariate time series forecasting in smart buildings based on feature selection. Energy Buildings **196**, 71–82 (2019). https://doi.org/10.1016/j.enbuild.2019.05.021
56. Yu, J., Kim, M., Bang, H.-C., Bae, S.-H., Kim, S.-J.: IoT as a applications: cloud-based building management systems for the internet of things. Multimedia Tools Appl. **75**, 14583–14596 (2016). https://doi.org/10.1007/s11042-015-2785-0
57. Cvitić, I., Peraković, D., Periša, M., Botica, M.: Novel approach for detection of IoT generated DDoS traffic. Wireless Netw. **27**, 1573–1586 (2021). https://doi.org/10.1007/s11276-019-02043-1
58. Chung, C.-M., Chen, C.-C., Shih, W.-P., Lin, T.-E., Yeh, R.-J., Wang, I.: Automated machine learning for Internet of Things. In: 2017 IEEE International Conference on Consumer Electronics—Taiwan (ICCE-TW), Taipei, Taiwan, IEEE, 6/12/2017, pp. 295–296
59. Ponce, H., Gutiérrez, S.: An indoor predicting climate conditions approach using Internet-of-Things and artificial hydrocarbon networks. Measurement **135**, 170–179 (2019). https://doi.org/10.1016/j.measurement.2018.11.043
60. Antonini, M., Vecchio, M., Antonelli, F., Ducange, P., Perera, C.: Smart audio sensors in the Internet of Things edge for anomaly detection. IEEE Access **6**, 67594–67610 (2018). https://doi.org/10.1109/ACCESS.2018.2877523

61. Solanas, A., Patsakis, C., Conti, M., Vlachos, I., Ramos, V., Falcone, F., Postolache, O., Perez-martinez, P., Pietro, R., Perrea, D., Martinez-Balleste, A.: Smart health: a context-aware health paradigm within smart cities. IEEE Commun. Mag. **52**, 74–81 (2014). https://doi.org/10.1109/MCOM.2014.6871673
62. Sayeed, M.A., Mohanty, S.P., Kougianos, E., Zaveri, H.P.: Neuro-detect: a machine learning-based fast and accurate seizure detection system in the IoMT. IEEE Trans. Consum. Electron. **65**, 359–368 (2019). https://doi.org/10.1109/TCE.2019.2917895
63. Zhang, Z., He, T., Zhu, M., Sun, Z., Shi, Q., Zhu, J., Dong, B., Yuce, M.R., Lee, C.: Deep learning-enabled triboelectric smart socks for IoT-based gait analysis and VR applications. npj Flex. Electron. **4**, 1–12 (2020). https://doi.org/10.1038/s41528-020-00092-7
64. Anuradha, M., Jayasankar, T., Prakash, N.B., Sikkandar, M.Y., Hemalakshmi, G.R., Bharatiraja, C., Britto, A.S.F.: IoT enabled cancer prediction system to enhance the authentication and security using cloud computing. Microprocess. Microsyst. **80,** 103301 (2021). https://doi.org/10.1016/j.micpro.2020.103301
65. Mauldin, T.R., Canby, M.E., Metsis, V., Ngu, A.H.H., Rivera, C.C.: SmartFall: A smartwatch-based fall detection system using deep learning. Sens. (Basel) **18**, 3363 (2018). https://doi.org/10.3390/s18103363
66. Fafoutis, X., Marchegiani, L., Elsts, A., Pope, J., Piechocki, R., Craddock, I.: Extending the battery lifetime of wearable sensors with embedded machine learning. In: 2018 IEEE 4th World Forum on Internet of Things (WF-IoT), Singapore, IEEE, 2/5/2018, pp. 269–274
67. Pathinarupothi, R.K., Durga, P., Rangan, E.S.: IoT-based smart edge for global health: remote monitoring with severity detection and alerts transmission. IEEE Internet Things J. **6**, 2449–2462 (2019). https://doi.org/10.1109/JIOT.2018.2870068
68. Barricelli, B.R., Casiraghi, E., Gliozzo, J., Petrini, A., Valtolina, S.: Human digital twin for fitness management. IEEE Access **8**, 26637–26664 (2020). https://doi.org/10.1109/ACCESS.2020.2971576
69. Chiu, M.-C., Ko, L.-W.: Develop a personalized intelligent music selection system based on heart rate variability and machine learning. Multimedia Tools Appl. **76**, 15607–15639 (2017). https://doi.org/10.1007/s11042-016-3860-x
70. Venkatesh, J., Aksanli, B., Chan, C.S., Akyurek, A.S., Rosing, T.S.: Modular and personalized smart health application design in a smart city environment. IEEE Internet Things J. **5**, 614–623 (2018). https://doi.org/10.1109/JIOT.2017.2712558
71. Farhangi, H.: The path of the smart grid. IEEE Power Energ. Mag. **8**, 18–28 (2010). https://doi.org/10.1109/MPE.2009.934876
72. Babar, M., Tariq, M.U., Jan, M.A.: Secure and resilient demand side management engine using machine learning for IoT-enabled smart grid. Sustain. Cities Soc. **62,** 102370 (2020). https://doi.org/10.1016/j.scs.2020.102370
73. Siryani, J., Tanju, B., Eveleigh, T.J.: A machine learning decision-support system improves the Internet of Things' smart meter operations. IEEE Internet Things J. **4**, 1056–1066 (2017). https://doi.org/10.1109/JIOT.2017.2722358
74. Elsisi, M., Mahmoud, K., Lehtonen, M., Darwish, M.M.F.: Reliable Industry 4.0 based on machine learning and IoT for analyzing, monitoring, and securing smart meters. Sensors (Basel) **21**, 487 (2021). https://doi.org/10.3390/s21020487.
75. Alhussein, M., Haider, S.I., Aurangzeb, K.: Microgrid-level energy management approach based on short-term forecasting of wind speed and solar irradiance. Energies **12**, 1487 (2019). https://doi.org/10.3390/en12081487
76. Tang, N., Mao, S., Wang, Y., Nelms, R.M.: Solar Power generation forecasting With a LASSO-based approach. IEEE Internet Things J. **5**, 1090–1099 (2018). https://doi.org/10.1109/JIOT.2018.2812155
77. Wang, Y., Shen, Y., Mao, S., Chen, X., Zou, H.: LASSO and LSTM integrated temporal model for short-term solar intensity forecasting. IEEE Internet Things J. **6**, 2933–2944 (2019). https://doi.org/10.1109/JIOT.2018.2877510
78. Liu, Y., Yang, C., Jiang, L., Xie, S., Zhang, Y.: Intelligent edge computing for IoT-based energy management in smart cities. IEEE Netw. **33**, 111–117 (2019). https://doi.org/10.1109/MNET.2019.1800254

79. Mirboland, M., Smarsly, K.: BIM-based description of intelligent transportation systems for roads. Infrastructures **6**, 51 (2021). https://doi.org/10.3390/infrastructures6040051
80. Awan, F.M., Saleem, Y., Minerva, R., Crespi, N.: A comparative analysis of machine/deep learning models for parking space availability prediction. Sens. (Basel) **20**, 322 (2020). https://doi.org/10.3390/s20010322
81. do Nascimento, N.M., de Lucena, C.J.P.: FIoT: an agent-based framework for self-adaptive and self-organizing applications based on the Internet of Things. Inf. Sci. **378**, 161–176 (2017). https://doi.org/10.1016/j.ins.2016.10.031
82. El-Wakeel, A.S., Li, J., Noureldin, A., Hassanein, H.S., Zorba, N.: Towards a practical crowd-sensing system for road surface conditions monitoring. IEEE Internet Things J. **5**, 4672–4685 (2018). https://doi.org/10.1109/JIOT.2018.2807408
83. Leung, C., Braun, P., Cuzzocrea, A.: AI-based sensor information fusion for supporting deep supervised learning. Sensors (Basel) **19**, 1345 (2019). https://doi.org/10.3390/s19061345
84. Dogru, N., Subasi, A.: Traffic accident detection using random forest classifier. In: 2018 15th Learning and Technology Conference (L&T), Jeddah, Saudi Arabia, IEEE, 2/25/2018, pp. 40–45
85. Chin, J., Callaghan, V., Lam, I.: Understanding and personalising smart city services using machine learning, The Internet-of-Things and Big Data. In: 2017 IEEE 26th International Symposium on Industrial Electronics (ISIE), Edinburgh, United Kingdom, IEEE, 6/19/2017, pp. 2050–2055
86. Ta-Shma, P., Akbar, A., Gerson-Golan, G., Hadash, G., Carrez, F., Moessner, K.: An ingestion and analytics architecture for IoT applied to smart city use cases. IEEE Internet Things J. **5**, 765–774 (2018). https://doi.org/10.1109/JIOT.2017.2722378
87. Haverkort, B.R., Zimmermann, A.: Smart Industry: How ICT will change the game! IEEE Internet Comput. **21**, 8–10 (2017). https://doi.org/10.1109/MIC.2017.22
88. Yun, J.P., Shin, W.C., Koo, G., Kim, M.S., Lee, C., Lee, S.J.: Automated defect inspection system for metal surfaces based on deep learning and data augmentation. J. Manuf. Syst. **55**, 317–324 (2020). https://doi.org/10.1016/j.jmsy.2020.03.009
89. Yang, J., Liu, M., Lu, J., Miao, Y., Hossain, M.A., Alhamid, M.F.: Botanical Internet of Things: toward smart indoor farming by connecting people, plant, data and clouds. Mobile Netw. Appl. **23**, 188–202 (2018). https://doi.org/10.1007/s11036-017-0930-x

Part I
Smart Environments

Chapter 2
Model-Based Digital Threads for Socio-Technical Systems

Marcus Vinicius Pereira Pessoa, Luís Ferreira Pires, João Luiz Rebelo Moreira, and Chunlong Wu

Abstract Smart environments can be built by connecting smart devices and control systems, which coexist as an integrated system that supports everyday activities. A digital thread gives the necessary support to introduce smartness in socio-technological systems. Digital threads are not only useful in the development of these systems, but also to facilitate the development of rich digital models or digital twins, bringing even more smartness to these systems. This chapter discusses how Model-Based System Engineering (MBSE) can be applied to design a digital thread of a Traffic Monitoring System (TMS) for a smart city. A TMS aims at increasing traffic safety by monitoring vehicles and pedestrians using road infrastructure, with potential impact on the reduction of environmental pollution and foster economic development. Designing a digital thread for a smart TMS is a challenging task that requires a consistent conceptual modelling approach and an appropriate design methodology supported by integrated tools. SysML (Systems Modeling Language) has been designed to support the specification of socio-technological (cyber-physical) systems like a TMS and gives integrated support to apply MBSE. This chapter shows how SysML can be applied to create a digital thread that defines the traceability between requirements, design, analysis, and testing. This digital thread represents both the physical and virtual entities of the system, enabling the development of digital twins for simulating, testing, monitoring, and/or maintaining the system. SysML is currently being redesigned, and the new SysML v2 aims to offer precise and expressive language capabilities to improve support to system specification,

M. V. P. Pessoa (✉) · L. F. Pires · J. L. R. Moreira
University of Twente, Enschede, The Netherlands
e-mail: m.v.pereirapessoa@utwente.nl

L. F. Pires
e-mail: l.ferreirapires@utwente.nl

J. L. R. Moreira
e-mail: j.luizrebelomoreira@utwente.nl

C. Wu
Hebei University of Technology, Tianjin, China
e-mail: wuchunlong@hebut.edu.cn

© The Author(s), under exclusive license to Springer Nature Switzerland AG 2022

G. Marques et al. (eds.), *Machine Learning for Smart Environments/Cities*,
Intelligent Systems Reference Library 121,
https://doi.org/10.1007/978-3-030-97516-6_2

analysis, design, verification, and validation. This chapter also discusses how SysML v2 is expected to facilitate the development of digital threads for socio-technological systems like a TMS.

Keywords Model-based systems engineering · Traffic monitoring systems · Smart environments · Digital thread · Digital twins

2.1 Introduction

With the introduction of interconnected smart Internet-of-Things (IoT) devices, it has been possible to create smart environments for different application domains, like, e.g., smart cities and smart industry. These smart environments have stringent requirements concerning the safety of their users, so that changes in these environments should be carefully considered before being applied. For example, in a smart city, the configuration of (autonomous) vehicles and infrastructure elements such as traffic lights and boards should be tested thoroughly before being modified, so that unwanted dangerous situations can be avoided.

To allow new versions or configurations of these systems to be tested before they are deployed, a digital representation of these systems can be defined, which is called a digital model [1–3]. In case this representation has a live connection with the actual system, which is possible by using sensors and actuators, it is called a digital twin [1–3]. The digital model (digital twin) should be as close as possible to the system it mimics, i.e., it should have high fidelity. This can be achieved by developing a digital thread that runs from the original system requirements to its design, implementation, and operation, providing traceability between these levels. This digital thread can be developed with Model-Based System Engineering (MBSE) techniques, from which SysML (Systems Modeling Language) is the most popular. SysML is a language for system design defined as an UML extension and supports the representation of requirements and various levels of detail of a system design and implementation.

In this chapter, we discuss the development of a digital thread to obtain a digital model (and ultimately a digital twin) by using SysML. A Traffic Monitoring System (TMS) example is used to illustrate this development, where the created digital model (digital twin) allows the TMS to be analyzed during operation, which may lead to a reconfiguration.

SysML is currently being redesigned and draft versions of the new SysML v2 standard-to-be are already available.[1] This chapter also discusses how SysML v2 is expected to facilitate the task of defining digital threads of socio-technological (cyber-physical) systems like a TMS.

The main contribution of this chapter is the definition of a MBSE-based methodology to define and maintain the digital thread of physical systems and their digital twins, properly illustrated with a case study.

[1] https://github.com/Systems-Modeling/SysML-v2-Release.

This chapter is further structured as follows: Sect. 2.2 introduces the concept of digital thread, discussing how it can lead to digital models or digital twins, Sect. 2.3 discusses the MBSE methodology considered in this chapter, Sect. 2.4 introduces our case study, which is a TMS that aims at monitoring an environment with vehicles and pedestrians by using road infrastructure and information systems, Sect. 2.5 presents the SysML models that we defined to illustrate the concept of digital thread, Sect. 2.6 discusses how SysML v2 is expected to facilitate the development of digital threads and Sect. 2.7 draws some conclusions and recommendations.

2.2 Digital Thread

This section introduces the concept of digital thread and discusses how a digital thread can lead to digital models and ultimately to digital twins of a target system.

2.2.1 Definition

The origin of the term *digital thread* can be traced back to a report on technological trends of the US Air Force [4], in which the digital thread was indicated as a technological game changer, together with digital twins. A digital thread was originally defined as 'the use of digital tools and representations for design, evaluation and life cycle management' [4].

To be effective, a digital thread should be supported by modelling tools that not only allow models to be produced in each phase of the lifecycle of the system, but also support traceability throughout the lifecycle. This is enforced by the term 'thread', which carries the connotation of a sequence of related (connected) items. For example, considering a simplified lifecycle with requirements capturing, high-level design, detailed design, implementation and operation, the modelling tools should allow the requirements described in requirements models to be related to the high-level design decisions and constructs of the high-level design and down to the implementation details of the implementation model. Some software tools are available in the market today that support this concept of digital threads (see, e.g., Syndeia[2] and Aras[3]).

Another connotation of the term digital thread that has been implicitly conveyed in [5], is to denote the 'integration of systems and communication channels that transfer digital twin data from sources to consumers', which focuses on how data flow in a system-of-systems containing a system on interest and its digital twin. Although this connotation is equally important, it is a completely different concept altogether, so confusion should be avoided.

[2] https://intercax.com/products/syndeia/.

[3] https://www.aras.com/en/why-aras/digital-thread.

2.2.2 Digital Representations

The definition of digital thread considered in this chapter refers to different representations of a target system, which should be made fit for some specific purpose. Considering, for example, the purpose of analyzing different configurations (versions) of the target system without any runtime connections to this system, this representation can be simply denoted as a *digital model* [1–3]. Tools can be used to perform static analysis or even simulate different versions of the target system until a satisfactory version is found, which can be deployed later.

Recent advances in sensing technology in the scope of the *Internet of Things* (IoT) have enabled the acquisition of large amounts of real-time data, which can be used to enrich the models of a digital thread. A *digital shadow* is defined as a digital representation that uses real-time data to fulfil its purpose. For example, a smart city may be able to sense the number of vehicles in circulation and its digital shadow may represent these data by plotting the vehicles on a map for the purpose of visualization.

2.2.3 Digital Twins

In addition to sensing, some mechanisms may be available that allow the digital representation to actuate on the actual system, for example, by changing its configuration or installing new software versions. A *digital twin* is defined as a digital representation that uses real-time data from the target system and has mechanisms to actuate on the target system at real-time [1–3]. Therefore, the digital twin can be considered as a specific digital representation in a digital thread, in which a real-time digital representation of a system is used to mimic the actual behavior of the system, for the purpose of analysis, behavior prediction, simulation, etc. Therefore, the digital twin can be obtained by combining the models produced in the digital thread with the data obtained from sensors, and actuating on the system itself, allowing possible corrective measures to be applied more quickly.

A digital twin is generally defined as a representation of a specific *system*, possibly but not necessarily at real-time, for some specific purpose (usage), like simulation, prediction, analysis, visualization, etc., so it is necessary to delimit this system. To fulfil the purpose of the digital twin, we may have to consider not only the system but also its (physical) *environment* and its *users* (agents). For example, in a digital twin of a plane, we may have to consider the air conditions (humidity, temperature, pressure) but also the pilot(s). We call the system, its environment, and users (agents) the *physical twin*, as the counterpart of the digital twin.

The digital twin requires *data* from the physical twin to fulfil its purpose. These data are obtained from the physical twin through IoT devices (*sensors*) that are connected to the digital twin through a network. In addition, the physical twin may give access to *usage data* (user interactions) and data on its *internal state* that is

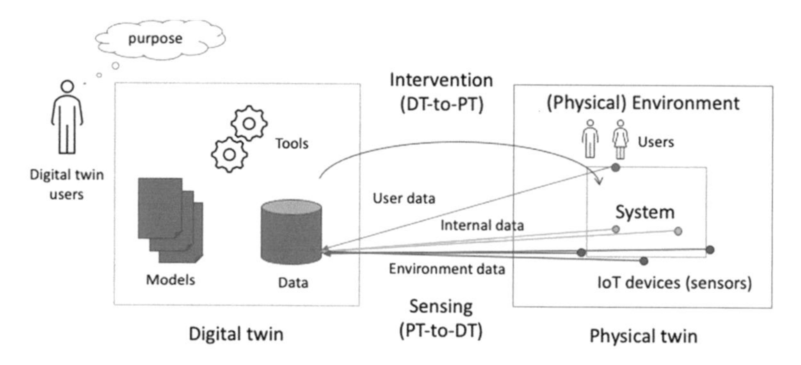

Fig. 2.1 Digital and physical twins

necessary for the digital twin to fulfil its purpose. A digital twin may be physically separated from or integrated with its physical twin, as discussed in [5].

Figure 2.1 shows a digital twin, its physical twin, and the elements necessary for them to operate properly.

2.2.4 *Digital Twin Dimensions*

Digital models, digital shadows, and digital twins are different types of digital representation that differ according to their data exchange. According to [5, 6], a digital representation contains up to five dimensions (see Fig. 2.2), and the digital twins are the most complete of these representations, as shown in Table 2.1.

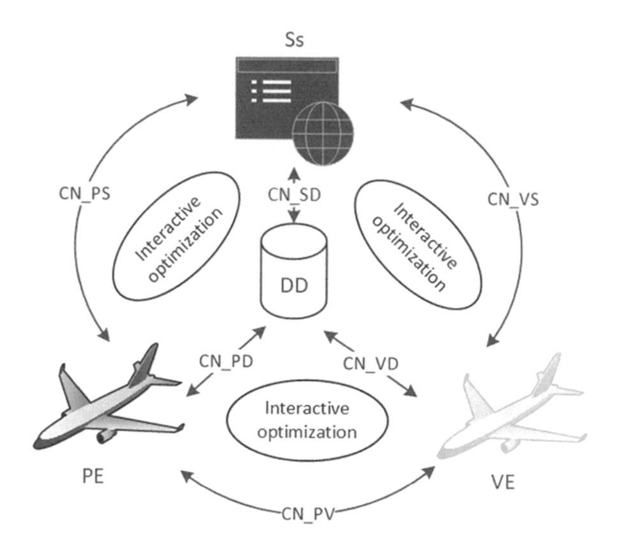

Fig. 2.2 The five dimensions of digital twins (adapted from [7])

Table 2.1 Digital representation and the 5-dimension framework

Dimension	Digital model	Digital shadow	Digital twin
Physical entity	X	X	X
Virtual entity	X	X	X
Services model	X	X	X
Data model	X	X	X
Connection model		X (PE → VE)	X (PE ↔ VE)

The *physical entity* (PE) or physical twin implements the basic functionality of the actual system and collects original data related to environments and operations. The *virtual entity* (VE) is a model of the element of the physical entity including structure, function, and environment, such as geometric and physical models. The *services model* (Ss) or tools include the business services for users and the functional services to support the digital twin itself. *Digital twin data* (DD) are driving factors of the model including data from other modules, knowledge, and fusion data. The *connection model* (CN) defines the connection between all the other dimensions, which allows real-time interactive communication between modules. These dimensions stress that by having only models to represent the reality, even if they are connected and exchange data with the physical twin, we still do not have a digital twin since services are necessary to fulfil to the purpose of the digital twin. These services are supported by the tools depicted in Fig. 2.1

A digital twin model represents the system according to its primary purpose. Normally a digital twin is created to estimate or keep track of the reliability, performance, or maintenance needs of the physical twin [5]. In this sense, the digital twin model does not necessarily have to represent the full real system, but only the system's aspects of interest. Therefore, this model must include both requirements, behavior, and structure from the real system and from the digital twin, as shown in Fig. 2.3. Knowledge of the requirements, behavior and structure of the digital twin is required for interfacing with the actual system, allowing the digital twin to perform performing its functions, which means that sometimes the hardware (i.e., sensors) or software of the physical twin (system under consideration) may have to be extended or adapted.

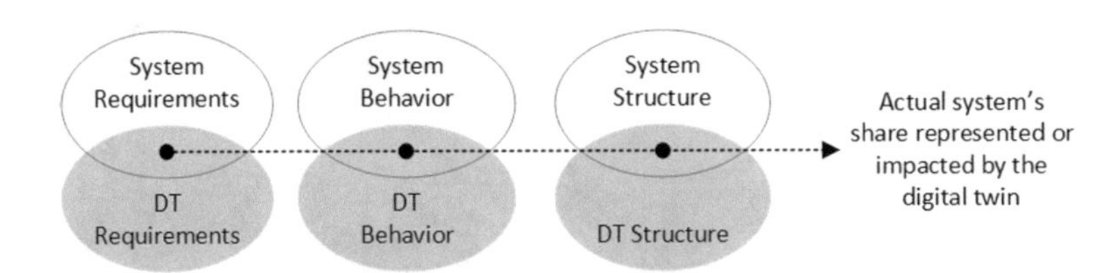

Fig. 2.3 Requirements, behavior, and structure of a digital twin and their impact with on the physical twin

2.3 Development Methodology

A digital thread can bring benefits to system development, and especially when introducing smartness in a socio-technological system like a smart city. However, a digital thread (as well as a digital twin) can be quite complex and expensive to produce. Furthermore, the exact purpose of a digital thread may not be known beforehand, which calls for generality and open-endedness [5]. By definition, a digital thread involves many representations of the system, related to different phases of a system life cycle, and produced by (and for) different stakeholders, so it is often multidisciplinary and rather complex.

2.3.1 Model-Driven System Engineering (MBSE)

INCOSE (International Council on Systems Engineering) defines *Model-Based System Engineering* (MBSE) as 'the formalized application of modeling to support system requirements, design, analysis, verification and validation activities beginning in the conceptual design phase and continuing throughout development and later life cycle phases' [8]. MBSE is a discipline to develop systems in which models play a central role from requirements engineering (generation and validation) to system synthesis (design and implementation) and system verification [5, 9]. In principle, MBSE does not prescribe any specific language, notation, or technique for the elaboration of these system models. However, when assuming that these models are applied to create a digital thread, this means that they should be integrated in a common framework so that they consistently can refer to different aspects of the system under consideration.

General-purpose languages like SysML and UML (Unified Modeling Language) have capabilities to represent structural and behavioral aspects of systems in an integrated form through interrelated diagrams, so they are natural candidates for being used to support MBSE. In this chapter, we concentrate on the use of SysML (Systems Modeling Language [10]), which is a graphical language for system design based on UML. SysML offers support to specify systems, in terms of their requirements, structure and behavior at different levels of abstractions. It is also quite common to use SysML to represent system structure and behavior in general terms, and use a more elaborate tool like, e.g., Simulink to perform a more detailed analysis by extending (enriching) the SysML model. SysML (structural) models in this case are used as the basis to provide an overview of the requirements and structure of the systems under consideration, while other techniques and their supporting tools can be used for more elaborate (specific) analysis tasks. For example, SysML models can be connected to models in, e.g., Simulink / MATLAB [11], CPN Tools [12] and SolidWorks [13], which give further details and analysis (simulation) capabilities to the SysML model elements.

In the literature, a digital twin is often defined as a physics-based, trustworthy model of the system [14], which can give the impression that SysML is not a suitable language for developing a digital twin. However, the main argument for applying MBSE and a general-purpose system design language as SysML, is that the digital twin is also a component in a (complex) system of systems, with its own digital thread. In [5], a development methodology for digital twins is defined based on MBSE, but this methodology is still neutral with respect to the languages that are used throughout the development life cycle. In this chapter, we show how some steps of such a methodology can be applied using SysML.

2.3.2 Models and Data

In Sect. 2, we stressed the importance of data when applying a digital thread to define digital twins, mainly because data can be related to the models of digital thread in different ways. The Models and Data (MODA) conceptual reference framework has been proposed in [15], which starts by identifying three different kinds of models, namely engineering models, scientific models and machine learning (ML) models, and defining their purposes as descriptive, predictive and descriptive, similarly to [16]. In addition, the authors have defined the different models and data can play in sociotechnical systems like the Traffic Management Systems we are considering in this chapter. Figure 2.4 shows the different types of models and data, and how they relate to each other in the scope of sociotechnical systems.

The relationships between models and data in Fig. 2.4 should be clarified before we can talk about the more specific roles played by models in the digital thread of a system like the TMS.

In a sociotechnical system, a descriptive model describes the sociotechnical system and/or the software system (e.g., the monitoring and control software). Data to produce these models are typically domain knowledge, in the case of an engineering model, but these could also be runtime data that can be used to tune (synchronize) the model with reality. A descriptive model can be transformed into a predictive model, for example, in case machine learning is used for analysis or prediction. We also assume that a descriptive model can be refined into a prescriptive model, for example, in a top-down development process. For the same reason, prescriptive models can be refined by adding implementation details. Assuming that software code is a special type of prescriptive model that prescribes what processors (or virtual machines) are supposed to do, software code can be seen as a prescriptive model that can be deployed on a software system. Finally, a prescriptive model can be enacted in a software system, for example, in case a digital representation of some spatial environment is rendered by a 3D software application. In this sense, a digital twin includes a descriptive model (virtual entity) and either (or both) prescriptive and predictive models (services), depending on the purpose of the digital twin.

Since a digital twin requires a descriptive model, in this chapter we concentrate on SysML, which is a language to represent both structural and behavioral aspects of

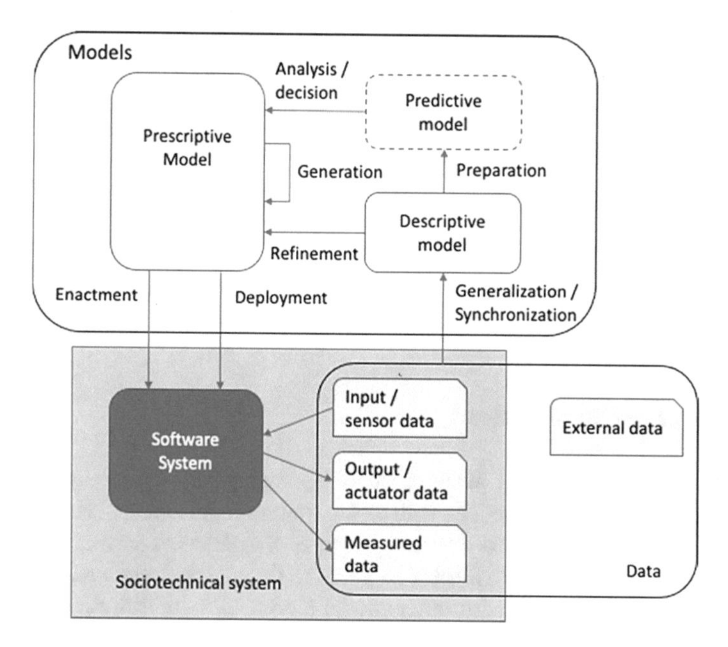

Fig. 2.4 Models and data relevant to a sociotechnical system (adapted from [15])

the system of interest. To draw a parallel, in conceptual modelling, many approaches use descriptive models to represent the multilevel and multi-perspective high-level abstraction of reality. Some of these approaches target the need for the evolution (changes) of such models, either because of the identification of semantic problems or because of the changes in the perception (point of view) of the designer. In this sense, the concept of an active conceptual model [17] can be compared to the concept of the digital thread, since the active conceptual model requires time and space modeling constructs, and should allow the traceability not only among the different model levels but also among the changes made (provenance), which can be a challenging problem for tracking objects (instances) that suffered changes over time.

2.3.3 *Life Cycle Support*

Figure 2.3 shows the life cycle phases that we are considering in this chapter, together with some concerns related to each phase, inspired by [5]. Below we briefly discuss the modeling needs of each of these phases, but in the remainder of this chapter, we concentrate on the first three phases, which are the most relevant to build the foundation for a digital thread. Although in [5] a top-down methodology is presented that is quite general, this methodology was mainly defined taking in mind a so-called 'fielded system' (in that case an unmanned surveillance vessel). In contrast, we target

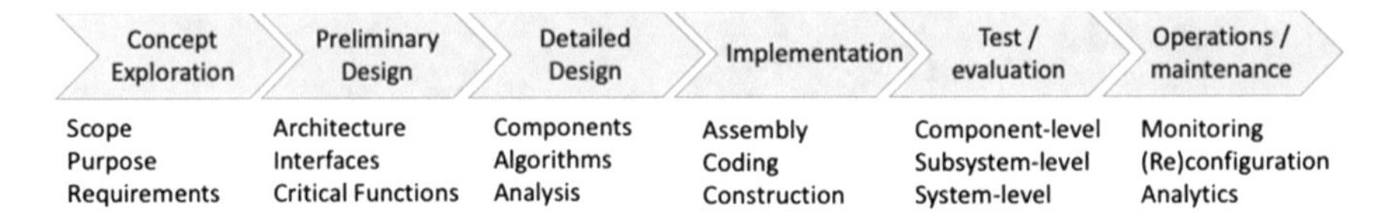

Fig. 2.5 System life cycle phases

a smart environment like a smart city, considering capabilities for traffic control. These differences are highlighted in the discussion below (Fig. 2.5).

2.3.3.1 Concept Exploration

In this phase, the scope and purpose of the system (or system-of-systems) being considered needs to be defined, as well as the system requirements. Although UML offers only use case diagrams for this purpose, SysML also supports a so-called Requirement Diagram, which allows its users to define requirements, relationships between requirements, and with other model elements. In addition, we can define use case diagrams and assign activity diagrams to them, to give more detail to the definition of the expected system behavior for some requirements. For a digital twin, the concept exploration phase aims at understanding the requirements, which lead to the definition of the system architecture in the preliminary design.

2.3.3.2 Preliminary Design

In this phase, the general structure of the system under consideration should be defined. In the case of a digital twin, the sensors and connections to the physical system should be defined in this phase at a high level of abstraction. In SysML, this can be accomplished with Block Definition Diagrams, but also with Internal Block Diagrams, which are meant to give more details about the internal structure of the high-level system components.

For traceability purposes, it is advisable to define how the requirements influence the modelling choices at this level, by defining relations between requirements and model elements of these block diagrams.

2.3.3.3 Detailed Design

In this phase, some more details of the system under consideration are given in terms of specific hardware and software models that have to be created. In some systems-of-systems, existing systems play the role of components and are integrated through well-defined interfaces, but other components may have to be implemented. A typical situation is in the development of a digital twin architecture, in which sensors generate data about a system or environment (e.g., in the case of a smart city), and a digital twin model must be defined and implemented to manipulate these data. Models produced in this step should also allow analysis, for example, through simulation.

2.3.3.4 Implementation

In this phase, the models produced so far are used as blueprints to create the concrete artefacts that will compose the final system of systems. For example, a digital twin design model may be implemented in a software system in a programming language like Python or Java, possibly making use of libraries for machine learning.

An interesting discussion is in case the system being considered has a software control component, like in a control room in a Traffic Management System. If this software component has been implemented in accordance with object-oriented techniques and properly using design models, these software components have an internal representation of the objects that they monitor and/or manipulate, which could be considered as a rudimentary form of the digital twin. The design models of these components could then be reused to accelerate the development of a digital twin for the environment under control, depending on the purpose of the digital thread.

2.3.3.5 Test and Evaluation

In this phase, the system is tested and evaluated. Models in this phase consist of test suites consisting of test cases, which can be automated and integrated with the development and deployment process like often used in DevOps. Test cases should relate to the (functional) requirements, making sure no requirement is left unverified.

In this phase, developers may have then the opportunity to start using the digital twin, collect test data, and train and evaluate the digital twin algorithms.

2.3.3.6 Operations and Maintenance

In this phase, the system is deployed and used by its end-users. If the purpose of the digital twin is to increase the availability of the system by improving its maintenance like in [5], in this phase the digital twin collects real-time data and processes these data, possibly using machine learning models, to advise the maintenance personnel about how to maintain the system.

A digital twin can also collect data in this phase that can be used in the earlier life cycle phases in a new iteration of the system development, in which new system releases and/or configurations are defined.

2.4 Traffic Monitoring System (TMS)

The case study of this chapter is a TMS, which is an information system that is an integral part of an Intelligent Transportation System (ITS) used in traffic management control centers by transportation agencies [18]. A TMS is a core system for mobility management that supports traffic analysis for the optimization of roadway system flows, predict transportation needs, and improve transportation safety. In this chapter, we consider a *smart TMS*, which is a key system to enable smart cities by using Internet-of-Things (IoT) technologies and big data through decentralized approaches for optimization of traffic on the roads, as well as intelligent algorithms to accurately manage traffic situations [19]. An excellent example of smart TMS can be found in the smart city initiatives in Hamburg (Germany) [20], where smart traffic lights provide data to the TMS, which reacts to congestion situations by offering routing solutions to control room operators and communicates with buses.

A smart TMS offers situational awareness through real-time efficient monitoring of traffic-related information, like the number of vehicles within road segments, the types of vehicles, their flow information (e.g., location and speed), and the road network infrastructure (e.g., traffic lights, segment direction and capacity, number of lanes and maximum speed). This information is obtained by collecting data from diverse sources, like road cameras, Eulerian and Lagrangian sensors, GPS systems, Intelligent Connected Vehicle systems, as well as non-conventional sources like social media streams and feeds [18]. Vehicular ad hoc Networks (VANETs) are quite promising to improve the efficiency of smart TMS. In VANETs, all interactions are supported by wireless communication links either between vehicles or between the vehicle and a roadside unit (or central server), where the communication technology follows the Wireless Access for Vehicular Environment (802.11p) standard, which uses channels at the 5.9 GHz band, one dedicated to the control channel, and the others dedicated to service channels [21].

One of the core capabilities of a smart TMS for improved traffic modeling is real-time vehicle classification to identify the vehicle types in the road, which is a key feature not only for traffic operation but also for transportation planning. For example, the geometric roadway design is based on the types of vehicles that frequently utilize the roadway, e.g., a highway section capacity and the respective pavement maintenance work plan can be estimated based on the number of large trucks in the segment [18]. Vehicle classification systems are usually categorized according to where the system is deployed: in-roadway-based, over-roadway-based, and side-roadway-based systems. Over-roadway-based vehicle classification can be implemented with the support of cameras, as well as infrared, ultrasonic and laser sensors on the ground, and UAV (Unmanned Aerial Vehicles) and satellite systems on the aerial space. The survey on intelligent TMS [18] also classifies the vehicle classification systems according to sensor types and how the sensor data are used.

The most common over-roadway-based vehicle classification system is based on surveillance cameras, which can provide rich information about visual features of passing vehicles and their geometry in multiple lanes. Digital image processing

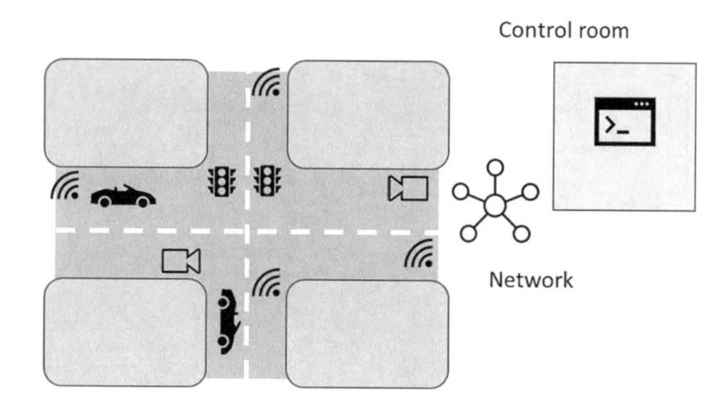

Fig. 2.6 Schematic representation of our Smart TMS

technologies together with adequate computational processing resources allow the classification of multiple vehicles in a quick and accurate way. For example, many approaches use surveillance cameras for vehicle classification and plate recognition [22, 23]. In short, a camera-based vehicle classification system captures the image of a car, extract features from this image, and executes an algorithm to perform the classification.

Most of the earlier versions of camera-based smart TMSs use simple classification models (e.g., decision trees) to extract features from the vehicle image, while more advanced approaches use machine learning techniques (e.g., deep learning) for more automatic and efficient classification methods. The survey [18] lists some of these machine learning approaches, comparing their accuracy, key features and which vehicle classes can be identified, like sedan, SUV (Sport Utility Vehicle), van, bus, truck, among others. Other vision-based methods that use different types of cameras besides video systems, e.g., omnidirectional cameras, aerial images, and closed-circuit television, are surveyed here [24], where neural networks are exploited for classification, training, and pattern recognition.

For the sake of simplicity, our case study on the TMS modelling consists of vehicles, roads, surveillance cameras, traffic lights, road sensors and a control system, as shown in Fig. 2.6. Furthermore, we assume that that traffic lights and road sensors are connected to the control system through a (wired or wireless) network, which is reliable enough to guarantee the reliable communication. In this fictional case, the main purpose of the digital twin is to help guarantee traffic flow performance. This simplified system has been defined to illustrate the concepts of digital thread and the use of SysML, which are discussed in Sect. 5.

2.5 The TMS Digital Twin Modeling Case

This section presents our case study, in which we discuss the digital thread of a TMS and aim at defining a digital twin of the TMS environment. In this section, we apply and refine the methodology presented in [5] by identifying the SysML diagrams that can be used in the development of digital twins and applying them to our case study.

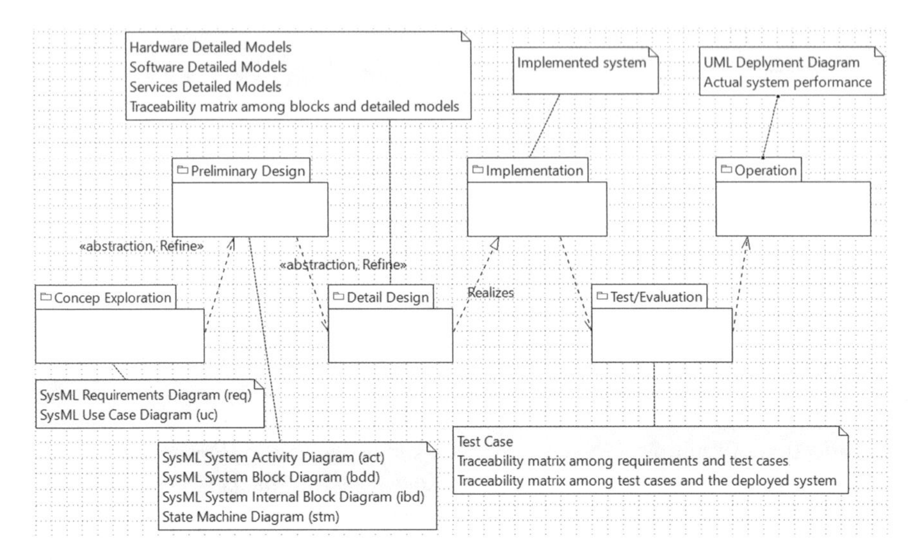

Fig. 2.7 MBSE methodology and related SysML diagrams and MBSE artefacts

Figure 2.7 presents a SysML package diagram, where each package corresponds to one phase of the methodology and contains the proposed diagrams and other MBSE artefacts for modelling a digital twin system of systems.

Since the system modeling effort and the use of SysML takes place mainly during the concept exploration and preliminary design, this section focuses on these two initial phases, because other MBSE techniques and the details on the hardware and software design are outside of our scope. Modeling was performed using Papyrus under Eclipse, namely Papyrus SysML 1.6 (4.8.0), Eclipse 2020–06 (4.16.0) and JDK 11.0.11.

2.5.1 Concept Exploration

Considering the digital twin's five dimensions, we assume that the SS (Services Model) and DD (Data Model) requirements determine the DT (Digital Twin) primary purpose and used data, the VE (Virtual Entity) requirements related to the chosen digital representation, the PE (Physical Entity) requirements include the subset of relevant requirements from the actual real/physical system, and the CN (Connection Model) requirements define the scope of the connections and their expected quality. These requirement diagrams together with a use-case diagram represent a generalized concept of operations (CONOPS), often used in (military) command and control systems. A CONOPS describes the characteristics of a proposed system from the viewpoint of the individual/actors who will use/interface that system.

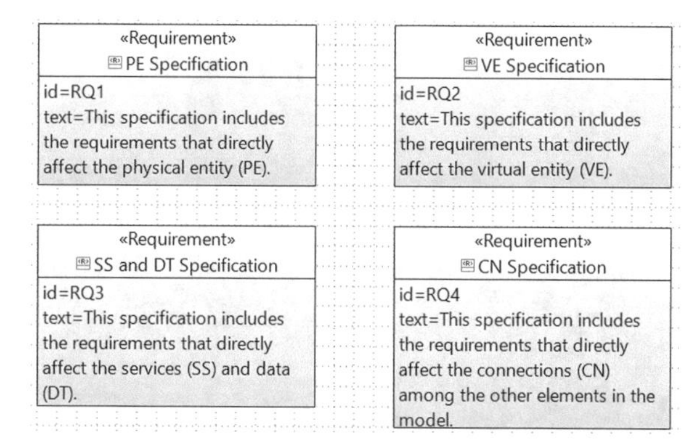

Fig. 2.8 Requirements' specifications

To organize the requirements set and facilitate their reuse, four specifications were defined for the PE, VE, SS + DD and CN, respectively. Once requirements in SysML are also containers, the specifications are themselves requirements (see Fig. 2.8), which have been further detailed in terms of linked and traceable requirements.

The SS requirements include the DT functionality necessary to fulfil the DT purpose. Since the TMS DT aims at managing and improving the traffic flow performance, the requirements define how this improvement takes place both at the tactical and the strategic levels, where the former deals with the current operations and the latter with the planning for further system improvements (see Fig. 2.9).

The PE requirements are a subset from the actual TMS and relate to the available sensors, displaying information and representing the actual road network (see Fig. 2.10). CN requirements set the expected quality related to the communication among the DT dimensions and include expectations in terms of reliability, availability, capacity, compliance, response time, etc. The implementation of the PE and CN requirements might lead to adding new or updating the actual system's hardware and software. The VE requirements detail the generic digital models' requirements. These generic and reusable requirements apply depending on the digital representation type. When modelling a DT, all the general requirements must be abstracted according to the specific scenario, which is the case of the TMS example in Fig. 2.11.

A use case diagram completes the concept exploration (see Fig. 2.12) and shows the relevant system actors (user and data sources) and the functions that allow their interaction with the system. All the functions must trace to at least one of the previously identified requirements. Figure 2.12 also shows the requirements of the "simulate flow alternatives" function.

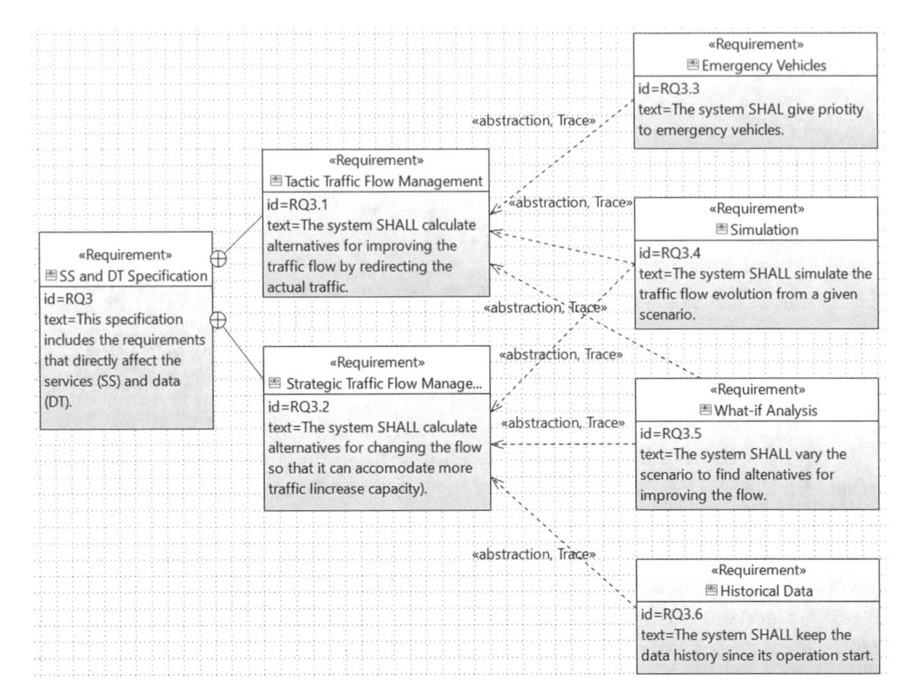

Fig. 2.9 Services model requirements

2.5.2 Preliminary Design

While the concept exploration phase focuses on understanding the system requirements and functions, the preliminary design defines the system architecture and identify the interfaces among the system components. To represent the system preliminary design, we show the following diagrams:

- A block definition diagram (bdd) that describes the system structure, which includes both hardware and software components.
- A block definition diagram (bdd) that describes the interfaces among the system components.
- An internal block diagram (ibd) for each complex system component.
- A state transition diagram (std) for each component controlled or monitored by the DT.
- An activity diagram (act) for each function identified in the use case diagram created during the concept exploration phase.

Figure 2.13 shows the system structure bdd in which the TMS is composed by four major components (road infrastructure, control center, vehicles, and network infrastructure) and their constituent components. In the TMS example, the VE is the digital model, which includes virtual representations from the circled components plus the road infrastructure.

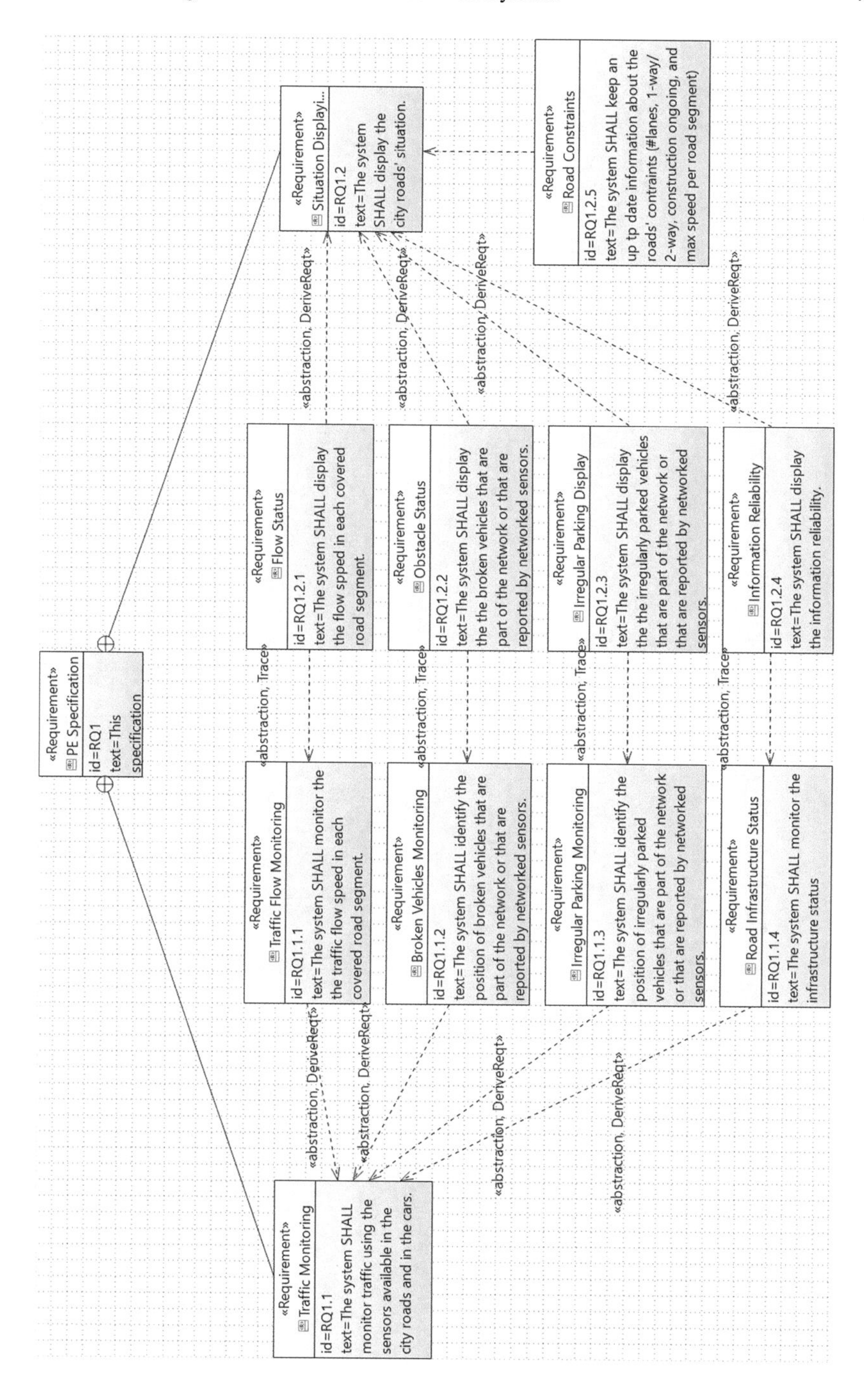

Fig. 2.10 PE model requirements

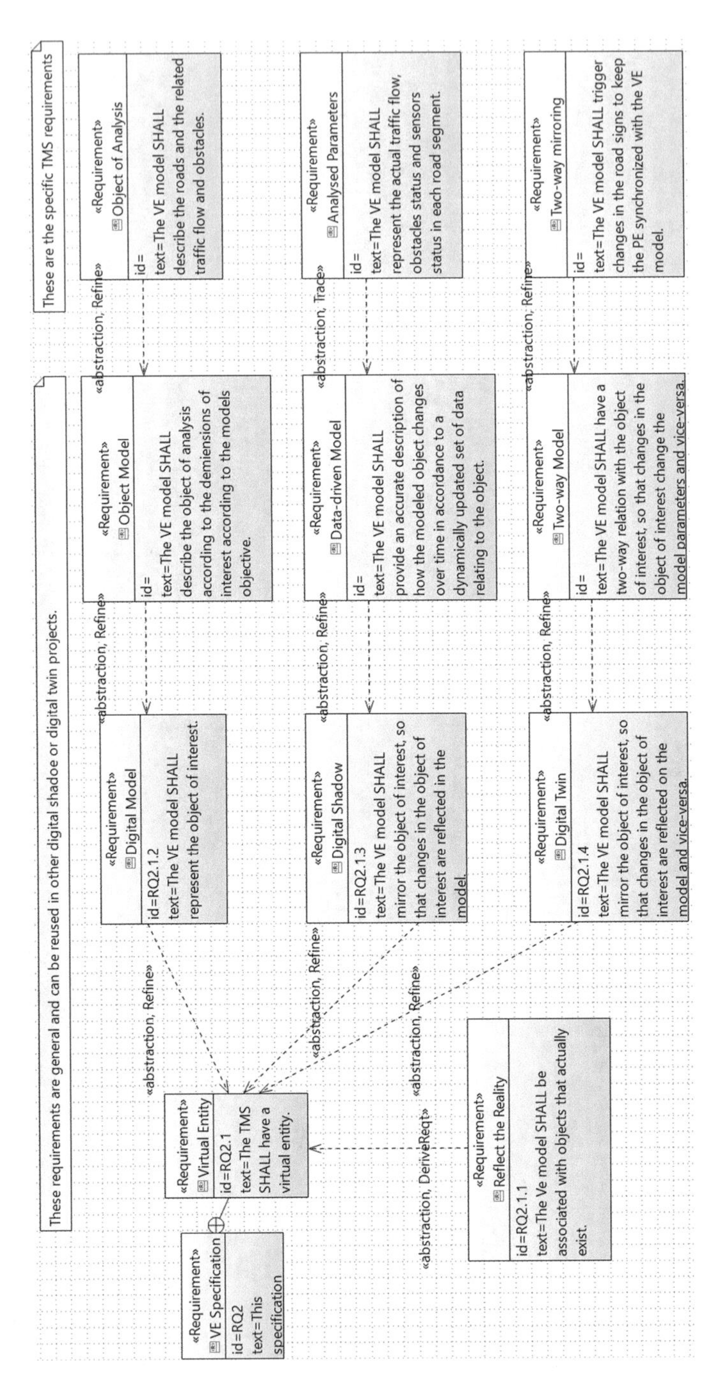

Fig. 2.11 VE model requirements

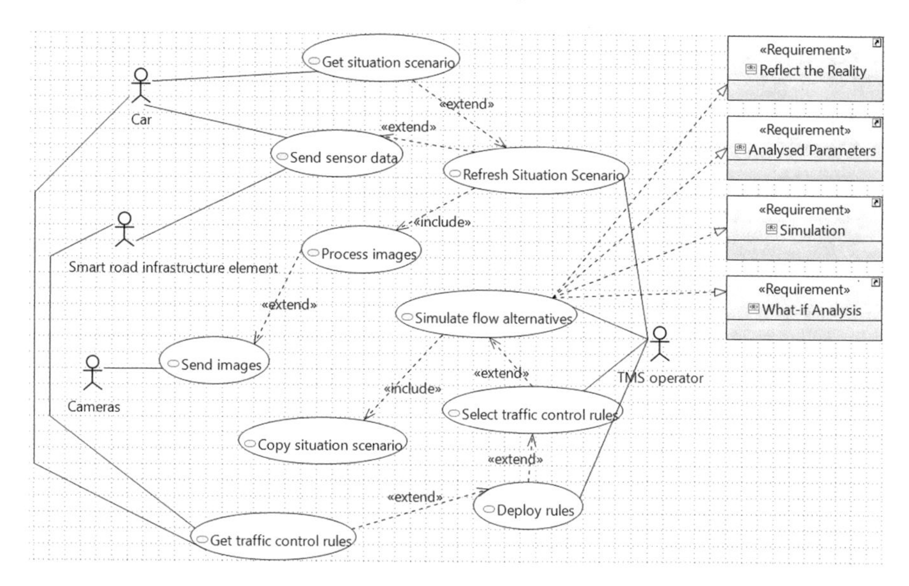

Fig. 2.12 System use-case diagram

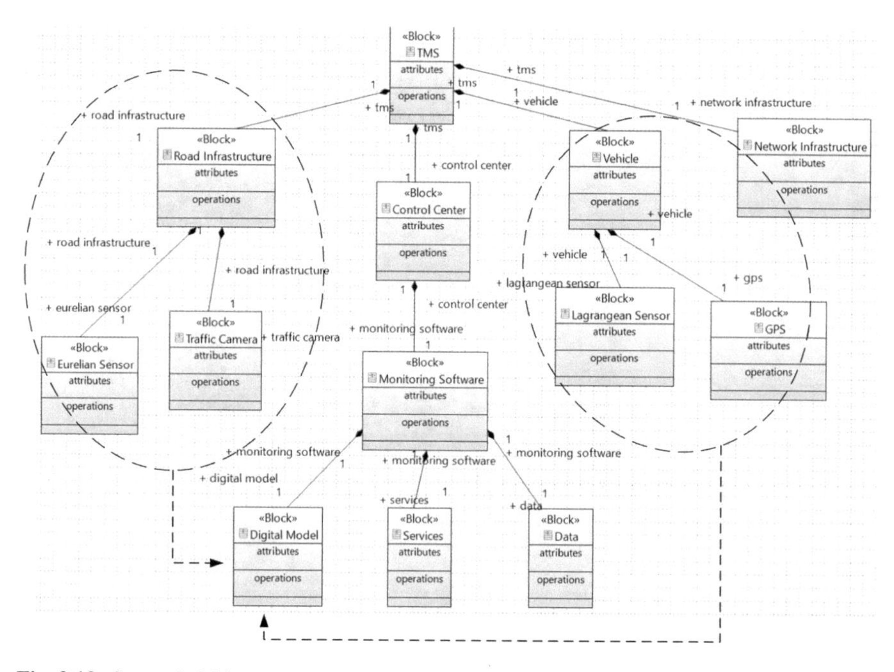

Fig. 2.13 System's bdd

Figure 2.14 includes the same blocks as Fig. 2.13 but shows how they interface with each other. The monitoring software modeled as a constituent part of the control center.

The VE includes the digital model internal elements. Figure 2.15 shows the need for synchronizing both the virtual vehicles and road structure elements with their physical counterparts.

State machines can help understand the behavior of the vehicles' and road structure's sensors (see Fig. 2.16). By combining these elements together with the road structure segments, a traffic model can be generated, which can be further used for simulating the traffic flow and finding better alternatives for absorbing the traffic (see Fig. 2.17).

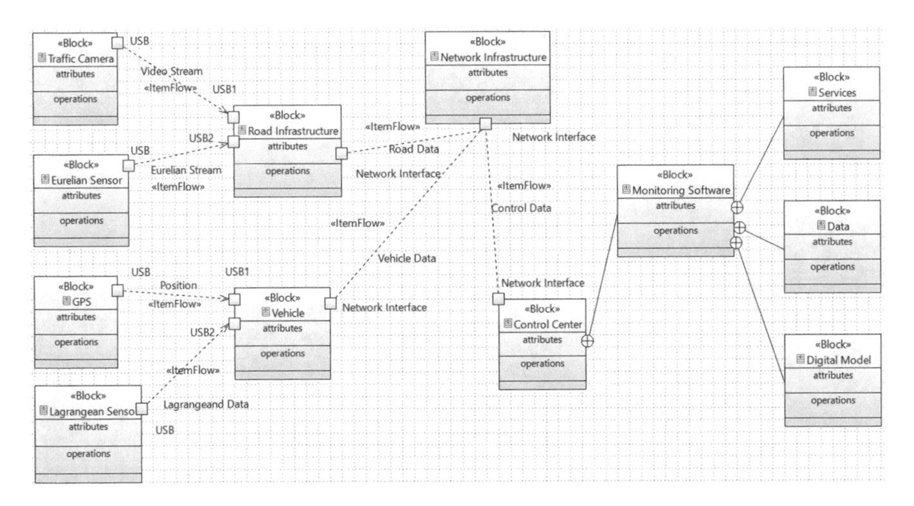

Fig. 2.14 System interfaces bdd

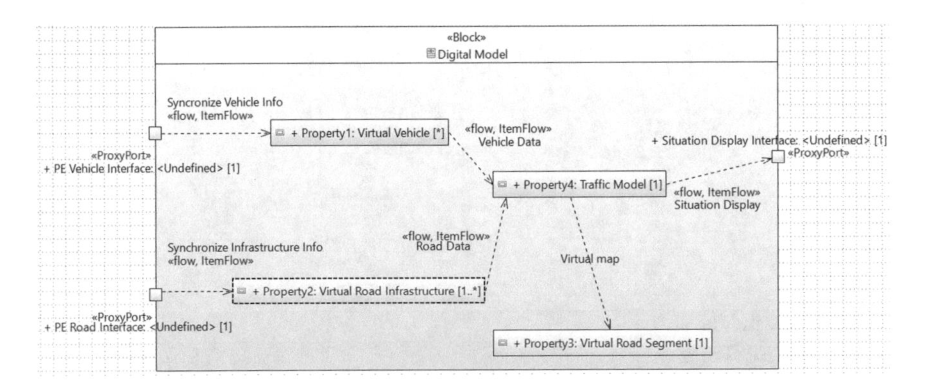

Fig. 2.15 Digital Model ibd

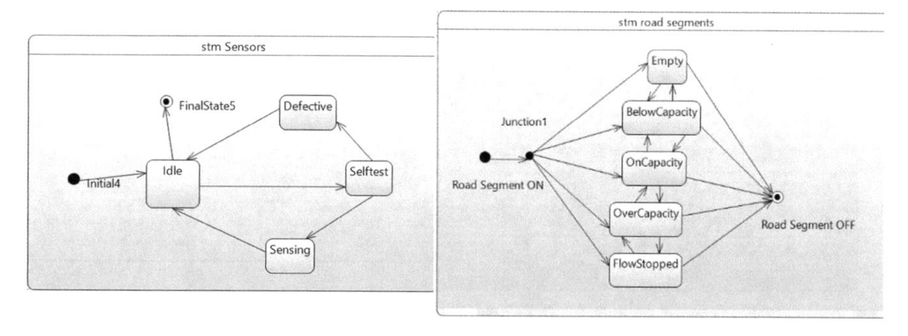

Fig. 2.16 State machines

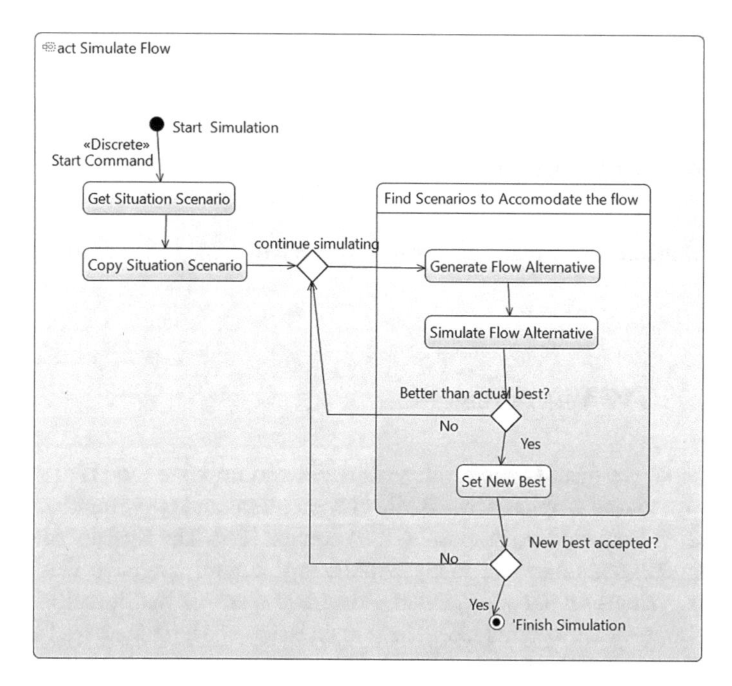

Fig. 2.17 Simulate flow activity diagram

2.5.3 Remaining Development Phases

During the remaining development phases the system is further detailed into more concrete hardware, software and service models (the latter in the case of product-service systems) and the implementation, testing and operations take place. In this way, a digital thread can be created with the proposed SysML diagrams and all related MBSE artefacts (see Fig. 2.18), The traceability among all elements allows the expected behavior (requirements) and the actual system performance to be compared,

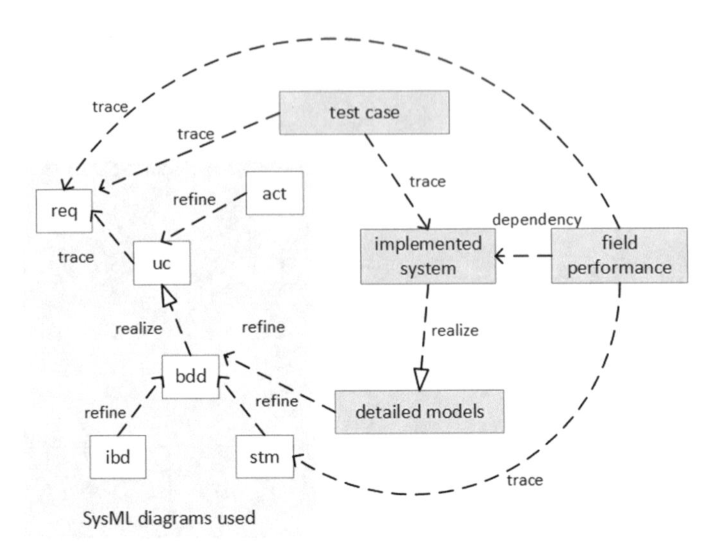

Fig. 2.18 MBSE and the digital thread

and these elements to be analyzed and reconfigured to come closer to the real system behavior.

2.6 SysML v2 Improvements

In this section, we discuss some future directions to improve the DT development methodology, giving emphasis to SysML v1.x problems and opportunities that are in the SysML v2 Request For Proposal (RFP) context [25]. The SysML development started over 20 years ago and today SysML is a mature language that is widely applied for systems engineering, having extensive research that identified its main issues and improvement points [26]. These main issues can be categorized as a formal foundation, metamodel, textual notation, and standard interface issues.

One relevant issue with the SysML 1.x formal foundation is that semantic checks on models can only be made manually by the designer after the specification, for example, by introducing OCL (Object Constraint Language) constraints to check semantic correctness, as usually done in UML. Instead of this approach, SysML v2 introduces automated validation checks during specification time, by formalizing the SysML language with a logic language that allows embedded semantic validation. In this way the designer can model, for example, a flow assuring that the units assigned to the ports are compatible, i.e., the types that are involved in the ports' connection can be automatically compared. With these definitions, the designer can automate the constraint checks of the connection, for example, between meter ports and kilogram ports.

In terms of metamodel technical issues, SysML 1.x is commonly implemented as a UML profile and, therefore, is limited by the UML embedded constraints. To address this issue, SysML v2 is being developed as a proper Domain Specific Language (DSL), i.e., following the Model-Driven Architecture (MDA) formalization practice with OMG (Object Management Group) MOF (Meta Object Facility) metamodel, making SysML v2 more flexible to address specific language problems. Furthermore, SysML 1.x has limited constructs to explicitly represent model variants. Variant modeling is an important technique to increase the variety in a family of products while minimizing development costs [27]. For example, the traffic modeling subsystem of a TMS is a variation point associated with a traffic model algorithm, which is a variation composed of different types of algorithms, where each algorithm is a variant and can be based on different vehicle classification models. This way, the DT design of the TMS could include the capability of comparing traffic models that use the output of in-roadway-based, over-roadway-based, side-roadway-based, and hybrid-based vehicle classification systems. In addition, variant modeling plays a major role in the design of TMS DT based on VANET technology due to its highly dynamic topology.

In addition to relieving SysML from the UML constraints and addressing variant modelling, by defining SysML v2 as a DSL appropriate textual and graphical notation can be defined in a systematic way. SysML v2 gives emphasis to the textual notation, adopting a clear syntax with emphasis on natural language, which is a trend in current modelling languages, like the case of Python for data science. This might be contradictory in SysML because it is usually associated with diagrams, but this choice allows the creation of alternative graphical notations based on the source model that is defined as text. This capability is demonstrated by the SysML v2 implementation reference with Eclipse and Jupyter Notebook.[4] In this way, different tools can implement the graphical rendering of the SysML v2 models according to their preferred notation, which is useful for domain-specific DT solutions, such as in the case of our TMS block diagram (Fig. 2.13), which could use a car icon for the Vehicle block, a frame icon for Eulerian sensor and a mobile icon for Lagrangian sensor.

DT solution providers can also benefit from the standardization of the SysML v2 Application Programming Interface (API), which adopts the Service-Oriented Architecture (SOA) principles and specifies a Platform Independent Model (PIM) that is used to define services and their operations to interact with the system and the models. For example, this approach allows the models to be queried to extract specific components and perform operations among different tools. Among different SysML v2 API implementation references, we highlight the API implementation based on RESTful web services that conform to OpenAPI specification and use a W3C semantic web standard (JSON-LD) to encode objects.[5] This capability is particularly interesting for DT solution providers to allow reliable and seamless integration

[4] https://github.com/Systems-Modeling/SysML-v2-Pilot-Implementation

[5] https://github.com/Systems-Modeling/SysML-v2-API-Services

of PLM (Product Lifecycle Management), CAD (Computer-Aided Design), requirements management and ALM (Application Lifecycle Management) tools, which is a new trend in Digital Thread platforms, like, for example, the Intercax Syndeia platform.[6]

2.7 Conclusion

This chapter brings further insights into the systematic development of digital threads to yield digital twin models based on SysML. Key concepts and practices are exploited, such as the difference among the concepts of the digital thread, digital model, digital shadow, as well as a digital and physical twin. These different types of digital representations are defined according to the connections among the physical and virtual entities, which vary from non-implemented, unidirectional and bidirectional. A digital twin is composed of five dimensions (physical entity, virtual entity, services, data and connections) and is more than just a set of models that are connected to a physical twin through IoT devices. A digital twin should also have a clear purpose and specific services to achieve their purpose, where these services can also determine the aspects and level of detail in which the physical entity needs to be represented in the virtual entity.

This study analyzes an existing MBSE methodology for digital twin development, highlighting the importance of how the three different types of models (prescriptive, descriptive, and predictive) can interact among them and with data from software systems. We investigated how a general-purpose systems design language like SysML can be used as a core standard to support this methodology, which only covers a high-level system engineering lifecycle support. The smart TMS case study was used to illustrate how this MBSE methodology with SysML is useful for the development of a digital thread towards a digital twin.

The results obtained from the TMS example confirm the insights reported in the literature and provide initial evidence of positive effects. For example, the SysML diagrams convey the necessary information for the concept exploration and preliminary design phases, which are leveraged by the reuse of the requirement definitions of the actual system performance. It is possible to compare the systems' actual results and initial requirements and trace how the different artefacts contributed to the result and get insight on improvement points to better fulfil the initial requirements. In addition, the reuse of these requirements is fostered by splitting the requirements according to the 5-dimension framework introduced here. We expect that the methodology presented in this chapter can be generalized and is applicable to other smart environments and that our case study provides enough guidance for elaborating the necessary models.

As a threat to validity, this research is limited since it provides one simplified example that covered only the first two development phases of the methodology.

---

[6] https://intercax.com/products/syndeia/

To obtain definite evidence, further work on the complete digital twin and digital thread development is required. Future work also includes the exploitation of the new features of SysML v2, which address some of the existing issues in SysML, such as the lack of formal semantics. SysML v2 also offers new opportunities to smart industry solution providers, such as the API for tools' integration. In this context, we foresee a new trend of digital thread platforms for digital twin development that are possibly based on MBSE methodologies, and implemented on top of IoT platforms, providing seamless integration of PLM and ALM with other engineering systems such as CAD, simulation, requirements management, and service management tools.

References

1. Fuller, A., Fan, Z., Day, C., Barlow, C.: digital twin: enabling technologies, challenges and open research. IEEE Access **8**, 108952–108971 (2020). https://doi.org/10.1109/ACCESS.2020.2998358
2. Kritzinger, W., Karner, M., Traar, G., Henjes, J., Sihn, W.: Digital twin in manufacturing: a categorical literature review and classification. IFAC-PapersOnLine **51**(11), 1016–1022 (2018, January 1). https://doi.org/10.1016/j.ifacol.2018.08.474
3. Oakes, B., et al.: Improving digital twin experience reports. In: Presented at the Proceedings of the 9th International Conference on Model-Driven Engineering and Software Development—MODELSWARD (2021)
4. Global Horizons: United States Air Force Global science and technology vision, appendix. United States Air Force Chief Scientist's Office (2013). [Online]. Available https://purl.fdlp.gov/GPO/gpo126317
5. Bickford, J., Van Bossuyt, D.L., Beery, P., Pollman, A.: Operationalizing digital twins through model-based systems engineering methods. Syst. Eng. **23**(6), 724–750 (2020). https://doi.org/10.1002/sys.21559
6. Tao, F., Zhang, M., Nee, A.Y.C.: Digital Twin Driven Smart Manufacturing. Academic Press (2019)
7. Wu, C., Zhou, Y., Pereira Pessoa, M.V., Peng, Q., Tan, R.: Conceptual digital twin modeling based on an integrated five-dimensional framework and TRIZ function model. J. Manuf. Syst. **58**, 79–93 (2021, January 1). https://doi.org/10.1016/j.jmsy.2020.07.006
8. Systems Engineering Vision 2020, INCOSE-TP-2004-004-02, International Council on Systems Engineering (2007)
9. Madni, A.M., Madni, C.C., Lucero, S.D.: Leveraging digital twin technology in model-based systems engineering. Systems **7**(1), 7 (2019). [Online]. Available https://www.mdpi.com/2079-8954/7/1/7
10. OMG System Modeling Language (OMG SysML) Version 1.6, formal/19–11–01, Object Management Group (2019). [Online]. Available https://www.omg.org/spec/SysML/1.6/
11. Chabibi, B., Douche, A., Anwar, A., Nassar, M.: Integrating SysML with simulation environments (Simulink) by model transformation approach. In: 2016 IEEE 25th International Conference on Enabling Technologies: Infrastructure for Collaborative Enterprises (WETICE), pp. 148–150 (2016, June 13–15). https://doi.org/10.1109/WETICE.2016.39.
12. Wang, R., Dagli, C. H.: An Executable system architecture approach to discrete events system modeling using SysML in conjunction with colored Petri Net. In: 2008 2nd Annual IEEE Systems Conference (2008, April 7–10), pp. 1–8. https://doi.org/10.1109/SYSTEMS.2008.4518997
13. Pietrusewicz, K.: Metamodelling for design of mechatronic and cyber-physical systems. Appl. Sci. **9**(3), 376 (2019). [Online]. Available https://www.mdpi.com/2076-3417/9/3/376

14. Wright, L., Davidson, S.: How to tell the difference between a model and a digital twin. Adv. Model. Simul. Eng. Sci. **7**(1), 13 (2020, March 11). https://doi.org/10.1186/s40323-020-00147-4
15. Combemale, B., et al.: A Hitchhiker's guide to model-driven engineering for data-centric systems. IEEE Softw. **38**(4), 71–84 (2021). https://doi.org/10.1109/MS.2020.2995125
16. Mavris, D.N., Balchanos, M.G., Pinon-Fischer, O.J., Sung, W.J.: Towards a digital thread-enabled framework for the analysis and design of intelligent systems. In: 2018 AIAA Information Systems-AIAA Infotech @ Aerospace (2018)
17. Spaccapietra, S., Parent, C., Zimányi, E.: Spatio-temporal and Multi-representation modeling: a contribution to active conceptual modeling. In: Active conceptual modeling of learning, Berlin, Heidelberg, Springer, Berlin, Heidelberg, pp. 194–205 (2007)
18. Won, M.: Intelligent traffic monitoring systems for vehicle classification: a survey. IEEE Access **8**, 73340–73358 (2020). https://doi.org/10.1109/ACCESS.2020.2987634
19. Sharif, A., Li, J., Khalil, M., Kumar, R., Sharif, M.I., Sharif, A.: Internet of things—smart traffic management system for smart cities using big data analytics. In: 2017 14th International Computer Conference on Wavelet Active Media Technology and Information Processing (ICCWAMTIP), pp. 281–284 (2017, December 15–17). https://doi.org/10.1109/ICCWAMTIP.2017.8301496
20. Späth, P., Knieling, J.: How EU-funded smart city experiments influence modes of planning for mobility: observations from Hamburg. Urban Transf. **2**(1), 2 (2020, December 31). https://doi.org/10.1186/s42854-020-0006-2
21. Akabane, A.T., Immich, R., Bittencourt, L.F., Madeira, E.R.M., Villas, L.A.: Towards a distributed and infrastructure-less vehicular traffic management system. Comput. Commun. **151**, 306–319 (2020, February 1). https://doi.org/10.1016/j.comcom.2020.01.002
22. Rudskoy, A., Ilin, I., Prokhorov, A.: Digital twins in the intelligent transport systems. Transp. Res. Procedia **54**, 927–935 (2021, January 1). https://doi.org/10.1016/j.trpro.2021.02.152
23. Wismans, L., de Romph, E., Friso, K., Zantema, K.: Real time traffic models, decision support for traffic management. Procedia Environ. Sci. 22, 220–235 (2014, January 1). https://doi.org/10.1016/j.proenv.2014.11.022.
24. Shokravi, H., Shokravi, H., Bakhary, N., Heidarrezaei, M., Rahimian Koloor, S.S., Petrů, M.: A review on vehicle classification and potential use of smart vehicle-assisted techniques. Sensors **20**(11), 3274 (2020). [Online]. Available https://www.mdpi.com/1424-8220/20/11/3274
25. Systems Modeling Language (SysML) v2—Request For Proposal (RFP), ad/2017–12–02, Object Management Group (2017)
26. Nigischer, C., Bougain, S., Riegler, R., Stanek, H. P., Grafinger, M.: Multi-domain simulation utilizing SysML: state of the art and future perspectives. Procedia CIRP 100, 319–324 (2021, January 1). https://doi.org/10.1016/j.procir.2021.05.073.
27. Colletti, R.A., Qamar, A., Nuesch, S.P., Paredis, C.J.J.: Best practice patterns for variant modeling of activities in model-based systems engineering. IEEE Syst. J. **14**(3), 4165–4175 (2020). https://doi.org/10.1109/JSYST.2019.2939246

Chapter 3
IoT Regulated Water Quality Prediction Through Machine Learning for Smart Environments

Ekleen Kaur

Abstract Smart cities have been contingent on the concept of Internet of Things since the beginning, the only measurable gap is the means to achieve it. Smart cities are not limited to urban housing but are more sectarian in the suburbs and require a means to devise efficient mechanisms and systems that can support sustainable growth. Industrial suburbs are the major pollutants that are the primary focus of this research. This paper works on the data of recycled wastewater procured from industrial use, which otherwise is directly discharged into the rivers. We collect the data using IoT sensors, these sensors are responsible for obtaining information about the data, also for scrutinizing and maintaining the water quality. The data contains major features that signify and influence the water quality, these parameters are used for the calculation of the water quality index. Finally, we train and predict this quality index using 3 machine learning algorithms-Decision Tree, Random Forest, and Deep Neural Networks. The evaluation metrics consist of RMSE, MAD, MAPE, and R squared value to identify and determine the model with the best performance.

Keywords IoT · Machine learning · Blockchain · Water quality index · Decision trees · Random forests · Deep neural networks

3.1 Introduction

This chapter is the extension of our previous work [1], in our last publication, we used blockchain technology as the key means to achieve an economic environment that can fundamentally support the functioning of sustainable development. The smooth functioning of our blockchain ecosystem is a major implementation that we seek to advance by using water quality prediction through machine learning. The aim of this research is to explicate such ways that deal with the problem statement in harmony with the blockchain-IoT relationship established in our previous implementation.

E. Kaur (✉)
University of Florida, Gainesville, USA

© The Author(s), under exclusive license to Springer Nature Switzerland AG 2022

G. Marques et al. (eds.), *Machine Learning for Smart Environments/Cities*, Intelligent Systems Reference Library 121,
https://doi.org/10.1007/978-3-030-97516-6_3

An Ethereum based Dapp has the provision of smart contracts [2] which allows the regulation of rules that are necessary to run a business. These rules are pivotal to our dapp as a lot of administrative policies are specified here, Ethereum uses a high-level language built exclusively for this-Solidity [3]. Solidity helps to achieve a cryptographically secured decentralised value transfer, with a custom specified set of rules. It is a peer-to-peer transaction-based immutable state machine that cannot change the existing data once it is cryptographically encrypted on a chain.

A general inference to the architecture of blockchain is the merkle tree directed acyclic graphs that is the most optimal definition of why the data on blockchain is immutable. In a simple merkle tree the present node stores the hash of its children nodes, the children nodes store their childrens' and so on. This behaviour concludes with the observation that any change in a node changes the hash value stored in the parent node which eventually affects the entire graph. Blockchain's property of immutable data is because of the same reason, every node is responsible for storing the hash of the previous block and any change in the chain results in the entire chain getting altered hence an illegal behaviour, which is not possible.

Ethereum Virtual Machine (EVM) supports the tokenization of assets, on variable standards ERC20, ERC721, ERC1155 and many more. The domain of this paper is on ERC20 for ethereum token exchange, trading a token on predefined benchmarks and with a calculable automated market maker model [4]. Tokens can be binded with a real-world asset and they act as a digital currency that can be regulated using a customized set of rules written on smart contracts. These assets can be any real-world entity encompassing a market value, the term sustainable asset is quoted for those assets that are interlinked with the environment and are renewable.

They can be Solar, Water, Hydro, Wind, Tidal or Geothermal, our scope in this chapter is water. Wastewater is an inevitable problem as organizational waste discharged into water results in the contamination of water bodies. The solution to this is a water treatment that already exists and has been adopted in various organizations that depend on water as a manufacturing product. The absence of a well-articulated regulation standard is a real-time problem that has led to an inefficient execution of water treatment.

IoT meters can be set up to work in the domain of regulation and inspection. They act as a quality keeper and can be deployed for obtaining live results in real-time. To consider the need for a security-keeper or a two-way authentication standard IoT helps to achieve this possibility. But water quality has some benchmarks and there are various categories that collectively are responsible for determining water quality.

Categorization of water parameters:

1. Temperature-Temperature might not directly affect the water quality but it is a significantly big factor affecting the variations inside other categories. Normal water temperature ranges from 20 to 27 °C.

2. pH-On a logarithmic scale pH can be defined as the estimate of the alkalinity of a solution on a scale of 1–10. Normal water has a pH value of 7. Industrial water is both acidic and alkaline in nature, Industrial water even after treatment might show some acidity or alkalinity because of the heavily acidic or basic effluents present before treatment.
3. Turbidity-Lose of transparency in water is the result of turbidity. Turbidity is the measure of the degree by which visibility is lost, measured in the Nephelometric Turbidity Unit (NTU). The presence of suspended particles in water is the cause of turbidity, the higher the particles in water the lower is the visibility and hence higher turbidity. Turbidity of drinking water is less than 5 NTU, for tap water it ranges from 5 to 25 NTU.
4. Biological Oxygen Demand-Bacteria are responsible for the decomposition of organic matter in water. So, by biological oxygen demand, we mean the amount of oxygen required by the bacteria which in turn are responsible for breaking the organic matter in water. Bacteria demand for oxygen is the quantified amount of oxidizable substances in water that contribute towards lowering the DO concentrations.
5. Fecal Coliform-The microorganisms present in water determine the quality of water and are their concentration in water is affected by the increase in temperature.
6. Dissolved Oxygen-Oxygen present in water is another evidence for good water quality, as oxygen is the key to sustenance for aquatic life. A desired DO content in water is essential for the continual dependency on aquatic life as a resource, in the case of Industrial water DO content helps analyze the standard for water quality.
7. Nitrate-They are an essential nutrient but in excess are the cause of aquatic problems, it is another benchmark to study the standard for water quality. In the case of industrial toxic pollutants, recycled water may have high nitrate contents depending upon the type of pollutant present in water. The high content of nitrate in water accelerates eutrophication and is also hazardous for human health.
8. Conductivity-Salinity is the cause of conductivity (ability to transfer current) in water. The number of salts present in water due to the acidic or basic nature of waste discharged into it varyingly affects the degree of electricity transferring capability of water.

These parameters are the prime consideration of IoT sensing in this chapter, the standard of water quality is judged cumulatively by all the categories combined. After an articulated analysis, the IoT sensing data is processed using machine learning techniques discussed in the later sections of the chapter. A basic discernment between the techniques used-

Decision Tree Algorithm-This algorithm splits the dataset to create smaller and smaller subsets of trees which eventually predicts the target value in the sample space. The tested feature is on the node and the list of possible outcomes are on the branches. The process is continued until no more new combinations can be generated hence the maximum depth of the tree has reached. Decision Tree works on various methods, ID3 is a very popular method for decision trees that creates a multiway tree. The method works on finding a categorical feature that can maximize the information gained (I_g) from the feature while also using impurity criteria i.e. entropy. In other words, information gain can be defined as the decrease in entropy (E) after the dataset is split on a particular feature (X).

$$I_g(T, X) = E(T) - E(T, X)$$

The problem with ID3 is that it is not applicable to numerical features, and our research uses numerical data. While ID3 is apropos for classification, CART, another Decision Tree method is fit for both classification and regression trees. CART, unlike ID3, creates a binary tree to find the best numerical feature by splitting the data on the basis of an impurity criterion. For classification, CART uses Gini impurity or towing criterion and for regression, it uses variance reduction.

Random Forest Algorithm-Random forests [5] are a type of decision tree, which work on tree predictors and their combinations. Trees are accurate with the data used to create them but the question of classifying new samples for prediction is variable. Through Random forests, the algorithm builds trees on values that are randomly sampled independently from featured vectors, at each step the algorithm considers a random subset of variables to build the tree.

Using bootstrapped data, that considers random variables at each step, results in assorted kinds of trees. The variety makes the algorithm more functionally better at performing certain tasks. For bagging, RF uses ensembling i.e. collection of all the predictions and results to make a final decision. For a given classifiers $c_1(x)$, $c_2(x)$, $c_3(x)_{\dots}$ $c_k(x)$, the binary random vector drawn for training-X, Y. The margin function can be defined as-

$$f(X, Y) = \text{avg}_k I(c_k(X) = Y) - \max_{j \neq y} \text{avg}_k I(c_k(X) = j)$$

Here I is the Indicator function, the margin function measures the difference of an average number of frequencies in the class of X & Y from the average number of frequencies in any other class. For a given probability P in the X, Y space, the generalization error is defined as,

$$Err^* = P(f(X, Y) < 0)$$

Deep Neural Networks-This algorithm is a type of artificial neural network which has multiple hidden layers between its inputs and the corresponding outputs. The activation function of DNN maps all the inputs from a lower layer, x_j, and it is transformed to a scalar state-y_j, this state is then fed to the upper layer.

$$y_j = f(x_j)$$

This x at any particular layer j can be defined as the sum of the bias at the unit j, with the summation of the products of all the outputs (y) at the lower layer i and the weighted connections (w) between both the layers i and j. For the purpose of activation, the function chosen is a sigmoid function with an output range [−1,1], represented as-

$$f\left(x_j\right) = 1/1 + e^{-xj}$$

where x_j can be defined as

$$x_j = b_j + \sum_i y_i w_{ij}$$

Logistic Regression is used to define the input–output mapping which has an output range [0,1]. All the weights and biases are pre-initialized prior to the training and later optimized by a cost function. This cost function is responsible for measuring the discrepancy between the predicted output with backpropagation and the target vectors. For the case of over-fitting of data the regularization term weight decay [6] is added to the cost function. Cost Function can be used for both classification and regression problems. Neural networks allow the computation of non-trivial problems by utilizing only a small number of nodes. For this purpose, we use DNN in the training of the information obtained through IoT sensors.

3.2 Literature Review

Machine learning bias on varied algorithms has different meanings, Professor Dietterich [7] analyzed the variance of decision tree algorithms and visualized the statistical bias of learning algorithms in his work. He proposed a coherent method to diagnose the problems related to machine learning bias and techniques to reduce variance. He used Breiman's bagging method to compare it with his work on tree randomization while voting decision trees. His conclusion on the perfect performance of the decision tree algorithm for letter recognition is tested in our research for a different circumstance using a different task and method of weighted bias. Navada [8] in his work overviews the most effective method in decision trees, he mentions ID3 as one of the oldest methods which are taken over by more efficacious algorithms like IDA (Intelligent Decision Tree Algorithm) and C4.5.

Segal [9] in his work on random forests pens down the conclusion that larger primary tuning parameter values affect the performance of the algorithm. Carrying his work forward in a special field of bioinformatics Yanjun [10] credits the increased use of this machine learning algorithm because of easily processed complex data structures, high-dimensional feature space, and small sample size. Itamar's [11] work on probabilistic random forests (PRF) proves the algorithm to be applicable for noisy data sets. His research improves the accuracy for noisy data sets by 10% using PRF which is useful for missing data values. Our analysis doesn't take noisy data in consideration, unlike in any of the previously mentioned works our data source is IOT sensors, so we discard the possibility of a missing categorical value in our research.

For problems dealing with frequentist statistical analysis, Bayesian analysis is a common approach [12]. Underpinning a powerful machine learning algorithm that uses Bayesian analysis, the Naive Bayes classifier [13] works on textual data. For our dataset on numeric values, we cannot use this technique as the algorithm cannot be used for training and predicting numeric values. Nugrahaeni [14] in her paper articulates the differences in the accuracy levels while using K-Nearest Neighbor (KNN), Support Vector Machine (SVM), and random forest (RF). She concludes that on a small set of data, SVM shows the highest accuracy, this result is gainsaid for larger sets of data where KNN and RF are proven to be better choices. Taylor's [15] article illustrates the use of KNN for estimating the total organic carbon (TOC) in sea water. His work is an essential factor for understanding TOC contents in seawater; amplifying these ideals, we consider the use of DNN for the IOT sensing data analysis, in hope of obtaining conclusive evidence that provides information about the nature of recycled industrial water from our trained dataset.

Convolutional Neural Networks (CNN) are computationally very intensive, this machine learning algorithm outputs the state-of-the-art results. The only problem with CNN is the huge amounts of storage memory required for reading weights, Avishek [16] introduces SRAM architecture that does not require reading of weights directly from the storage. In his work, he proposes a solution to the heavy energy requirement during the computation of binary weight network (BWN) which is a technique under CNN.

Qian [17] experiments on the parametric TTS synthesis using DNN, his work emphasizes the state boundary information which is essential for yielding a better-synthesized speech using DNN. Previously, DNN has been efficiently trained for various domains like speaker recognition [18], cloud computing [19], large vocabulary speech recognition [20], verbal emotion recognition [21] but so far on sustainable assets the applicability of DNN remains nature specific. Since our work is the first research on applying machine learning algorithms on IoT sensing data consisting of recycled waste water. Our research takes into consideration some of the aforementioned learning algorithms like RF, Decision tree (DT) and DNN to critically analyze the accuracy of each algorithm on our dataset. These prediction algorithms support and improvise the problems with IoT in terms of security, scalability and quality monitoring for our business network that is proposed for ethereum blockchain.

The varied compatibility of IoT services allows it to have a common feature with blockchain, i.e. scalability. One of the most commonplace confrontations in IoT is the need for smarter security. Elngar [22] specifies a genetic algorithmic (GA) approach in areas of healthcare. The evolution of GA in the weight calculation of Artificial Neural Networks (ANN) has a significant effect on tamper detection in healthcare applications. Other researchers such as Ogu [23], Samarth [24] have been concerned about theft detection for electricity and energy using IOT concepts. Samarth's [24] work expounds the IoT implementation for preventing energy tampering in sensing, Ogu [23] contributes in the physical inspection of infrared sensors in the case of human intervention.

Tampering leads to changes in the channel state information (CSI) [25], Bagci's work focuses on analyzing this CSI fluctuation at multiple receivers for human movement-related observation and its differentiation from tampering. In real-time the study of tampering varies from scenario to scenario, Zhang's [26] study is limited to a device and data-dependent verification scheme. The physical layer authentication is taken to be device-specific in his work; the scheme uses a dynamic key to generate a tag which is later embedded at the time of information transmission with the help of a hiding scheme, decoded at the receiver's end.

Numerous obfuscation strategies exist for tamper detection; many are domain-specific and efficient, others might not be domain-specific but are not that efficacious. In Sung Ryoung's [27] work they use anti-reverse engineering by employing steganography, the security checks are contingent on their test application. Breitenbacher's [28] work on Linux-based IoT applications in his host-based anomaly detection of IoT Malware has 100% efficiency as claimed by the researcher but has the prerequisite of memory requirement which causes CPU load. The possibility of a large number of parallel running IoT applications might be a big causality in the case of heavy CPU load. One of the common drawbacks of all the above works is that they don't elucidate the cross-domain compatibility of their research.

IoT for smart tracking devices [29], smart health networks [30], smart energy monitoring [31] has already been explicated by scientific researchers. The idea for IoT to be cross-domain compatible and easily integrable by not being platform-specific is still nebulous. The need arises with the drawbacks in IoT itself as IoT can offer services but for real-time solutions, IoT should be compatible with other technical interfaces. In this research work, blockchain is the major concern in terms of cross-domain compatibility. We elucidate the network of IoT with blockchain and highlight the major issues in the integration of both technologies.

Through machine learning, we plan to predict the model accuracies for various learning algorithms and train the dataset for calculating the absolute error in the prediction of the values [32–34]. By using different algorithms for the same dataset we can understand the categorical variation which is contingent on various factors and hence can derive the most optimal variant which provides the best performance. This chapter aims to mitigate the problems arriving while deploying the IoT sensors in real-time, through machine learning.

In our proposed business model, the domain was recycled wastewater; none of the aforementioned works so far have taken into consideration the use of sustainable assets to be tokenized as a currency of trade, as discussed in our previous work [1]. According to Mazzei [35], blockchain tokenizers for IoT applications can be used for supply chain by distributed management, she highlights the challenging task of integrating industrial IoT on blockchain platforms. In accordance with IoT security, Jea-Min [36] proposes blockchain-based firmware verification in her article. The milestone of maintaining IoT services on the blockchain network needs a smart approach, where the solution can learn and train itself from the previous IoT sensing data. Our approach through machine learning for the same concern provides many conclusive pieces of evidence from the trained model. We analyzed the statistical variation of the different categories of water and the influence of the independent variable on each one of them.

3.3 Dataset Description

We published the dataset on Kaggle [37], for the interpretation of the data the following categories of water quality are recorded using IOT sensors-Temperature (measured in °C), pH, Turbidity (measured in NTU), Biological Oxygen Demand (measured in mg/l), Total Fecal Coliform (measured in MPN/100 ml), Dissolved Oxygen (mg/l), Nitrate (mg/l) and Electrical Conductivity (measured in μmhos/cm). These categories in the dataset are impart in the calculation of the water quality index [38].

WQI is an important and efficient method to determine the degree to which the water can be put to use, a good water quality corresponds to drinking use and the lower level determines urban and industrial use cases. Our dataset is regarding the treated industrial wastewater, so catechizing the water quality does not aim at a prime WQI but rather the aim is to find the degree of reusability by calculating this benchmark. Information related to the categorical range variation and the data concentration within the range is elaborated in the next section.

3.4 Methodology

Summarizing the work in the previous paper [1], our problem case was about intensifying the usage of recycled water which ultimately de-escalates the dependency on fresh water for industrial use. The blockchain prototype of the wastewater tokenization carefully outlines a cogent set of rules that were about the governance, regulation standard, issuing of WRC tokens, token exchange, minting of tokens under specified circumstances and the transfer of tokens from one partaker in the business network to another.

The issuing of tokens was on the basis of percentage reuse of recycled waste water, this percentage reuse was the result of probing the IoT sensing data taken at defined intervals throughout the day. Since the token price is highly contingent on the numerical value which eventually outputs the number of partakers passing the quality probe and the numbers of partakers otherwise, a smart learning algorithm was needed to understand the fluctuations in the future data which might steer a serious impact on the token price. To delve into the intricacies of this research we filter the data obtained from the IoT sensors, train and test it to derive the predicted categorical variation and calculate the absolute error in the prediction results.

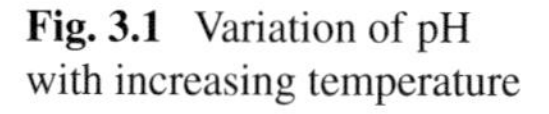

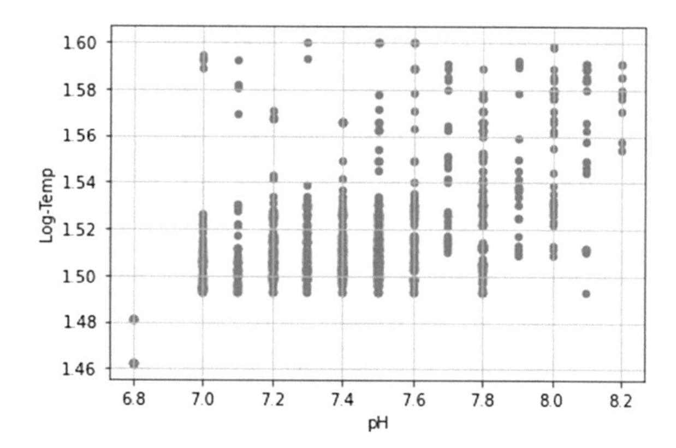

Fig. 3.1 Variation of pH with increasing temperature

Previously the purpose of a regulator [39] in the blockchain network was to monitor the water quality, for this, the paper proposed IoT sensors to act as an automated regulator. The consideration of minimizing man power now expands to the need for a smarter approach to regulation. For the selection of a particular machine learning algorithm where each algorithm is intrinsic to the kind of data and remains problem-specific, our selection is guided by the performance of each algorithm on our data. For our given data on recycled water quality parameters, we train it on some of the popular learning algorithms like DT, RF and DNN.

Our recycled water data is marginally alkaline in nature so the pH range is generally > 7–7.5 up to 8.2, however nearly 1.8% of the data has a pH value equivalent to 6.8, the rest is either neutral or alkaline as shown in Fig. 3.1. B.O.D level is usually < 5 mg/l for drinking water, our readings report the B.O.D range for treated wastewater to be between 31–51 mg/l. According to WHO, good water quality has turbidity ranging from 0–5 NTU, for drinking purposes the range changes to 0–25 NTU, in the case of our data the turbidity ranges from 42–50 NTU. In our data, total fecal coliform ranges between 4700–5650 MPN/100 ml; dissolved oxygen ranges from 3 to 4 mg/l, this category quantifies between 6.5–8 mg/l in the case of drinking water. For nitrates, our data ranges between 0.10 and 0.30 mg/l and for electrical conductivity, it is 200–400 μmhos/cm.

Temperature is a category that doesn't affect the water quality index directly. Even though WQI is not directly affected, temperature affects all the other categories as they vary in a certain way with the contradistinction in temperature hence it does influence WQI. For instance, turbidity can increase temperature, so for a higher temperature reading an increment in the turbidity might be the responsible cause, similar relationship is observed for conductivity as well it increases with the increase in temperature. Other features such as dissolved oxygen have an inverse relationship with the temperature, as colder water is observed to have a higher concentration of oxygen. So, for the quantified estimation of WQI, we don't use temperature as a feature but while during the training of our dataset we include temperature as one of the feature vectors. For the skewness of the large data, the logarithmic value of temperature is taken to graphically estimate the categorical variation of other features

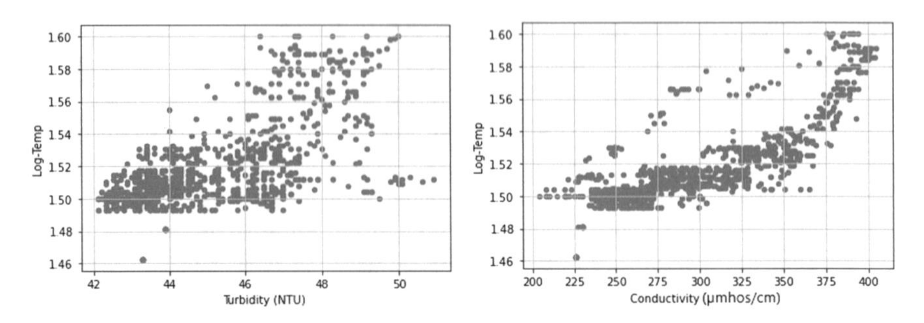

Fig. 3.2 Variation of turbidity and conductivity with increasing temperature

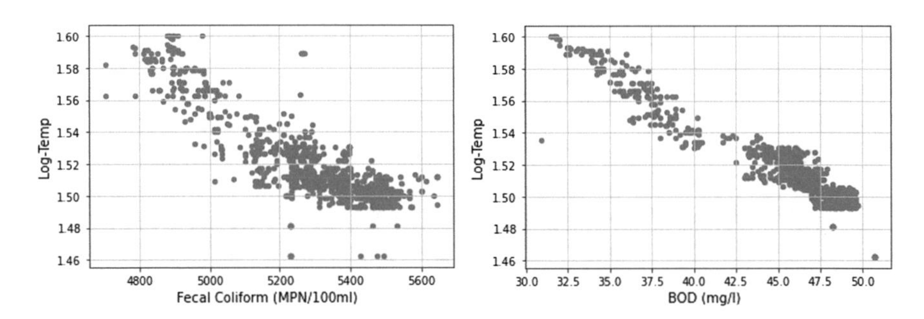

Fig. 3.3 Variation of fecal coliform and BOD with increasing temperature

according to temperature. Figure 3.2 shows the turbidity and electrical conductivity varying according to temperature.

It is quite evident from the data that turbidity and electrical conductivity have an uptrend with the rise in temperature, however, a major concentration of the recorded readings are concentrated between 42–46 NTU and 230–340 μmhos/cm. Figure 3.3 shows the categorical variation of total fecal coliform and B.O.D with respect to temperature. Showing higher concentrations at lower temperatures, the data is majorly congregated between 5230–5560 MPN/100 ml and 44–50 mg/l for the coliform count and oxygen demand respectively.

A downtrend of dissolved oxygen at higher temperatures and the predominance of observations between 3.8–4.0 mg/l signifies a lot about the water quality of the data. The mass predominance of nitrates between the range 0.1–0.3 mg/l is more at a higher value of temperature, so, there is a linear relationship observed between the two categories, however, a lot of unspecified intricacies and external factors may lead to a variation in real-time (Fig. 3.4).

Our dataset (D) uses regression, D consists of various readings for all the categories of water ranging from $K \in \{1,2,3 \ldots, N\}$, these categories map the inputs set: $\{i_1^k, i_2^k, i_3^k \ldots i_n^k\}$ to the output target: $\{o^k\}$. The error estimation of the prediction for the K inputs is given by $err_k = o_k - \hat{o}_k$, where $\hat{o}_k$ is the predicted value, and o_k is the calculated value. For regression, there are various ways to estimate the error in

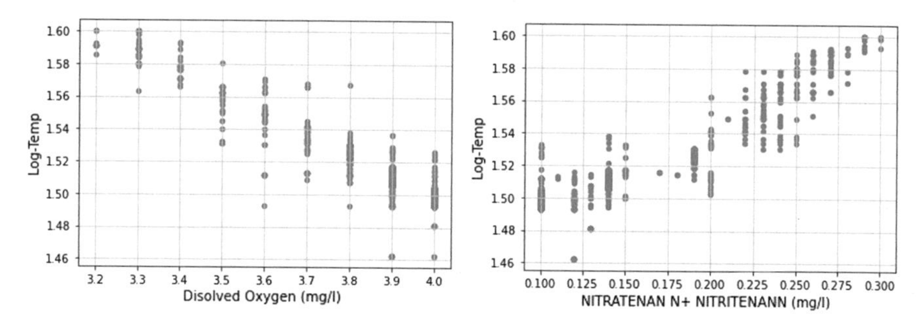

Fig. 3.4 Variation of dissolved oxygen and N + with increasing temperature

the prediction. Classification has an accuracy score but for regression, we define the better performance of the model using the error estimation. The evaluation metrics at which the errors in the predetermining evaluation were calculated for ML are-

Mean Absolute Deviation (MAD)-Converting the error recorded for each observation after obtaining the difference between the predicted value from its actual value, to positive, in order to obtain the absolute numeric value of the data and then dividing the sum of this difference for all observations from the total number of observations gives MAD. Mathematically it can be represented as-

$$\text{MAD} = \left[\sum_{i=1}^{n} \text{abs}\left(\text{err}^{k}\right)/\text{n}\right]$$

$$\Rightarrow \left[\sum_{i=1}^{n} o_k - \hat{o}_k\right] \Big/ n$$

Root Mean Square Error (RMSE)-RMSE is the quadratic mean of the difference, the square root of the residuals divided by the total number of observations. While measuring RMSE, it aggregates the magnitudes of all the errors recorded in the data points during the prediction to be encapsulated into one single measure of predictive power. RMSE is a measure of accuracy that assesses the forecasting errors within the dataset.

$$\text{RMSE} = \left[\sum_{i=1}^{n} (\text{err}_k)^2\right]^{1/2} \Big/ n$$

$$\Rightarrow \left[\sum_{i=1}^{n} \left|o_k - \hat{o}_k\right|^2\right]^{1/2} \Big/ n$$

Both MAD and RMSE are scale-dependent error estimation metrics.

Mean Absolute Percentage Error (MAPE)-Percentage error prediction of forecasting on the dataset can be defined as the summation of the absolute difference between the predicted and actual readings divided by the actual reading of the data.

This sum numeric of the absolute difference is then divided by the total number of observations in the data, for percentage this quantity is multiplied by 100. The advantage of MAPE is that it is scale-independent and is used for finding out relative measures.

$$\mathrm{MAPE} = \left[\sum_{i=1}^{n} |err_k|/o_k\right] * 100/n$$
$$\Rightarrow \left[\sum_{i=1}^{n} \left|o_k - \hat{o}_k\right|/o_k\right] * 100/n$$

The above three metrics are the standard used for error estimation in this research, we compare the above standards for all the algorithms used for the training of sensor data. The last benchmark for the analysis of the model is the REC curve [40]. They are an improvised version of the Receiver Operating Characteristic (ROC) curve, in ROC the area under the curve determines the probabilistic measure of the expected performance of classification. The complexity of the curve represents automatic unbiasedness, for 2 class use cases, the ROC curve is constructed by varying the threshold and then plotting the false positive and true positive on the X and Y-axis respectively with the curve lines interpolating between them. REC represents Regression Error Characteristics, they use efficacious strategies for representing the results to non-experts, compared to RMSE, REC curves output the results in a much more compelling manner [41]. The equation of the linear regression model can be represented as:

$$O_k = I_k \beta + Q,$$

Here, O_k is the [n × 1] vector of result observations, I_k is the full rank order of the design input matrix [n x q], β is the vector-matrix of unknown parameters, and ϙ is the vector of arbitrary errors. The accuracy for the REC curve increases with the increase in ϙ. For ROC the measure of error estimation is given by the Area under the curve similarly it is Area Over the Curve (AOC) for the regression error analysis of the REC curve.

Training our data on multi-regression models is a classical approach [41], even though it is easy to interpret the problem of the ability to learn only through linear mapping is consistent. Tree structures are better substitutes in this regard, beginning with decision trees, while modeling our data on this algorithm the hierarchical structure has an assorted set of rules, these if-then rules help determine the ultimate feature vector whose value is to be predicted in our case the water quality index [42]. This chapter uses 40% of the dataset for testing the three algorithms.

3.5 Results and Discussion

In the feature selection during the modeling of decision trees, the criterion to measure the quality of the split was selected as mean squared error, due to better training output it was modified to mean squared error with Friedman's improvement score. The threshold value of the model for early stopping is set to 0, the node will split till it reaches the leaf. The minimum quantified sample requirement while splitting the internal node is 2. The minimum number of samples that are required at the leaf node is set to 1. Unlike the splitting of nodes till the point only pure leaves are left which is less than the minimum number of sample splits, the maximum depth of the tree is set to 10. For the scheme used for the splitting of each node both the strategies-best split and random split were tested but on careful scrutiny, the final model was trained by choosing the best split. Cost complexity pruning is not set for the model, and the parameter is on the default value 0.

The graphical error estimation using a decision tree graph shows the following observations in the prediction of quality index. The major calculable results of water quality index having zero error is around 53.8, 56.7, 57.2 and 57.4. Out of the 1000 trained observations on DT a majority of about 956 readings don't forecast an erroneous prediction. The remaining frequency of positive and negative errors have a preponderance towards positive i.e. positive errors have a larger weight than negative. After adjusting the weight parameters to mitigate the class imbalance the histogram shows a nearly equable frequency of the positive and negative errors for the remaining 44 observations.

While training our model using Random Forests, due to bootstrapping of the data we don't use the entire 60% of the dataset for building each tree. The minimum impurity decrease is changed from default 0.0 to 0.1, reusing the solutions of the previous call, is set to false. The parameter for multi-output is set to average the errors of all outputs with uniform weight. The minimum number of samples at the leaf node is set to 1 with the minimum sample split of the internal node at 2, each tree is split till the leaf and the threshold value for early stopping is zero. Maximum features selection using random forest regression is equal to the number of features and the number of trees in the forest for the WQI prediction is 100.

The error figures using random forests are more articulate in outlying prediction characteristics and have a lesser range variation than the previous model. The quality index prediction reading at 53.8 and 56.7 nearly has a zero error, 57.2 has a marginal positive error and the concentration of quality index at 57.4 has a significant negative error. Other than the error distribution for frequency analysis, we refer to the histogram for RF, where we observe a discernible shorter error range in this model as compared to DT which was [− 2, + 2], refer to Fig. 3.5. From Fig. 3.6, the error range for RF is notably observed to be within [− 0.15, + 2.2], however, the frequency of this range distribution is around 964 readings which are at zero error, out of the remaining 15–24 readings are within the inner range [− 0.05, + 0.05] and the residual figure estimate is between the inner and outer range. For evidently

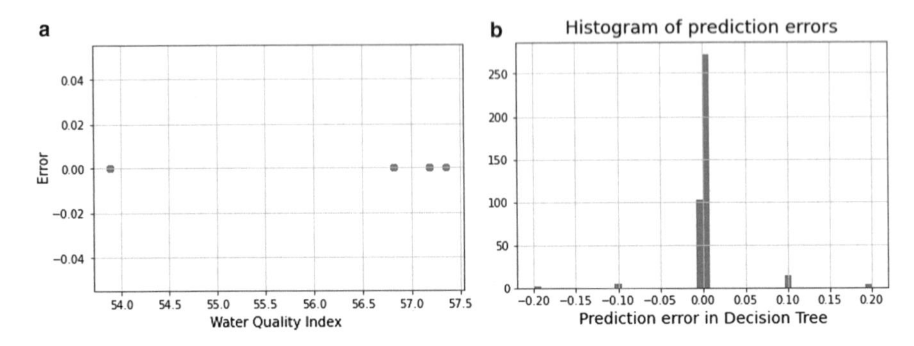

Fig. 3.5 **a** Distribution of erroneous predictions of WQI using DT. **b** Distribution of the frequency of reported errors

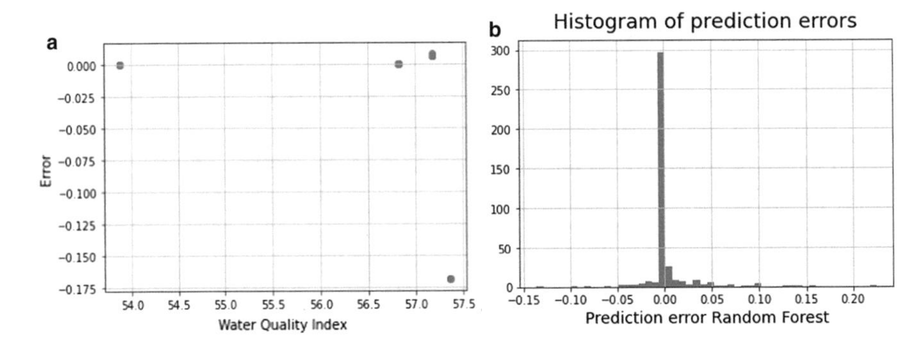

Fig. 3.6 **a** Distribution of erroneous predictions of WQI using RF. **b** Distribution of the frequency of reported errors

understanding the differences we calculate the error value on the three metrics for all the three training algorithms, refer to Table 3.1.

Deep Neural Networks are nonlinear functions, this human brain-inspired algorithm has various popular architectures, multilayer perceptron being one of them. This research considers this architecture which has three hidden layers each containing H hidden nodes along with a logistic function. The perceptron has one output node with a linear function. The neural network is adjusted using 3 hidden layers, with 100 epochs of the Broyden–Fletcher–Goldfarb–Shanno (BFGS) algorithm. For the activation function we use Scaled Exponential Linear Unit (SELU), for the first two layers of activation, the last layer uses Exponential Linear Unit (ELU). To prevent the

Table 3.1 Reported observations on three evaluation metrics

Algorithm	RMSE	MAD	MAPE
Decision tree	3.46e−04	8.09e−05	6.4e−03
Random forest	3.34e−04	1.14e−04	7.38e−03
Deep neural network	2.64e−02	1.73e−02	1.124

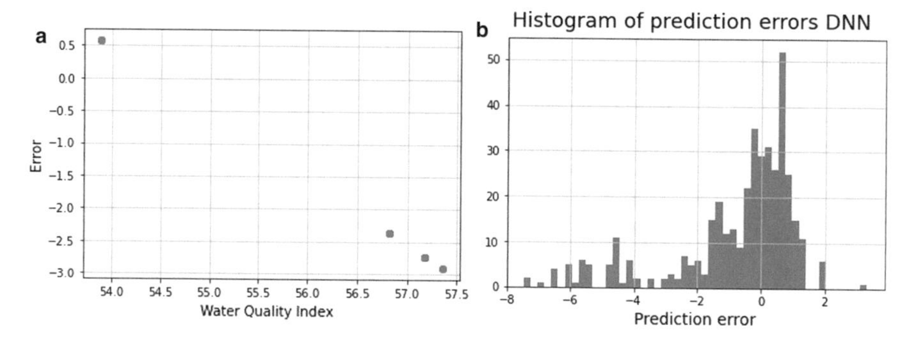

Fig. 3.7 **a** Distribution of erroneous predictions of WQI using DNN. **b** Distribution of the frequency of reported errors

neurons from synchronizing to their own weights and creating a bias we prevent them from converging to the same goal by decorrelating the weights using 30% dropout.

The number of neurons used for the training of the data before the activation of each hidden layer is 100, 100, 60, 1. The outputs are post-processed using the inverse of logarithm transform, in the case of negative results, the results are set to zero. To scrutinize the forecasting performance using DNN, a thirty run of a tenfold equating to a total of 300 simulations was applied to each tested configuration, along with an internal tenfold grid search on the training data. This tenfold grid search is used to find the best values of $H \in \{2, 4, 6, 8, 10\}$, which is then used to restrain the training data (Fig. 3.7).

The error estimation for DNN is higher than the other two models. For the positive error in data, the histogram shows that it has a higher frequency of occurrences than the negative errors, the quality index reading at 53.8 has a positive error of 0.5. The data points 56.7, 57.2 and 57.4 have a higher value of error below zero, the frequency of negative errors in the data might be less but the range varies till -8 and for the positive errors the same ranges up to $+3$. Out of the 1000 observations, 884 are nearly at zero error and the remaining shows a variable erroneous forecasting within the range $(-8, +3]$. Aberrations in data prediction using DNN are observed to be higher than in RF. To comprehend the model characteristics for better statistical analysis we report the error statistics on the following three metrics RMSE, MAD and MAPE which can be referred to in Table 3.1.

The error results reported on the three metrics clearly show that RF is the model with the least reported error. The final evaluation metric in order to deterministically conclude the predictive model with the best performance is the REC curve. REC is a representation method under ROC, it is a comprehensive illustration of the diagnostic ability of a regressor on a graphical plot as the discrimination threshold is varied. In order to observe the error distribution throughout the output range we use the REC curve to visualize the error in percentage of correct prediction with respect to the absolute tolerance in the prediction of water quality index, Fig. 3.8 shows the REC curve of all three models in this research. The r squared value which is the statistical measure representing the variance for a dependent variable (WQI) with respect to an

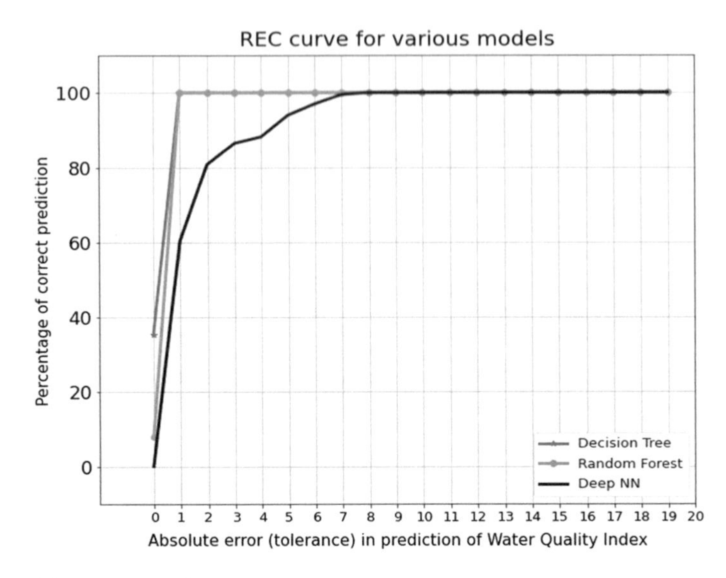

Fig. 3.8 REC curve for various models

Table 3.2 Reported reading of R^2 score

Algorithm	R^2 score
Decision Tree	9.98e−01
Random Forest	9.99e−01
Deep Neural Network	− 1.45

independent variable (other feature vectors), can be estimated using the ratio of the AOC of our REC curve on estimating the percentage of correct forecasting on WQI, to the AOC for the null model. Refer to Table 3.2 for the r squared value for each of the curves.

We observe the highest error count for all the metrics reported in DNN, for the RF and DT the error count of observed readings in RF is much better than DT. Even though both the models relatively have the same R squared value, the RF predictor is built by averaging outputs which result in a substantial improvement over DT.

To deterministically identify the best performance of the learning algorithm for our nature specific dataset, it is responsible for affecting a lot of intricacies in the business model exemplified in the previous paper.

- Having a learning algorithm provides a value estimate for short periods on the monthly percentage reuse of wastewater, this quantified numerically directly affects the token price.
- The best performance algorithm helps to discern and comprehend the WRC token variation.
- The analysis of the market cap for the token.
- Framing of exclusive rights for the liquidity providers

- This helps in the understanding of the liquidity pooling of tokens.
- Finally, through this study, we can contrive the rules for establishing the market maker model for our blockchain ecosystem [21].

3.6 Conclusion

In this chapter, we marked the asset training of wastewater through machine learning. Post visualizing the phylum according to the variations with respect to temperature, we enumerated the water quality index from the data containing the standard marking characteristics of recycled water. After a holistic view of reported results, we concluded that out of the three proposed models, random forest is the best training algorithm for our nature specific data procured through IoT sensing. Given the tolerable range of minimal errors in the prediction of WQI, random forest has demonstrated the best results out of the three.

The purpose of this research serves as a base architecture for our future works that conjecture the automated market maker model in our ecosystem. By analyzing the factors affecting the token price in this research, we plan to scheme the theorizing of the liquidity pooling for partaker organizations. In our future contributions, we even plan to brainstorm the authentication of the partaker organizations, an efficient mechanism of achieving this would be through zero-knowledge proof.

References

1. Kaur, E., Oza, A.: Blockchain-based multi-organization taxonomy for smart cities. SN Appl. Sci. **2**, 440 (2020). https://doi.org/10.1007/s42452-020-2187-4
2. Udokwu, C., Kormiltsyn, A., Thangalimodzi, K., Norta, A.: An exploration of blockchain enabled smart-contracts application in the enterprise (2018). https://doi.org/10.13140/rg.2.2.36464.97287
3. Gavin, W.: Ethereum: a secure decentralised generalised transaction ledger Istanbul version. https://ethereum.github.io/yellowpaper/paper.pdf
4. Karp, H., et al.: Nexus mutual: a peer-to-peer discretionary mutual on Ethereum blockchain. https://nexusmutual.io/assets/docs/nmx_white_paperv2_3.pdf
5. Breiman, L.: Random forests. Mach. Learn. **45**(1), 5–32 (2001)
6. Krogh, A., Hertz, J.A.: A simple weight decay can improve generalization. In: Moody, J.E., Hanson, S.J., Lippmann, P.R. (Eds.), Advance in neural information processing systems-4. Morgan Kauffmann Publishers, San Mateo, pp. 950–957 (1992)
7. Dietterich, T.G., Kong, E.B.: Machine learning bias, statistical bias, and statistical variance of decision tree algorithms. Technical report, Department of Computer Science, Oregon State University (1995). http://citeseerx.ist.psu.edu/viewdoc/download?rep=rep1&type=pdf&doi=10.1.1.38.2702
8. Navada, A., Ansari, A.N., Patil, S., Sonkamble, B.A.: Overview of use of decision tree algorithms in machine learning. IEEE Control Syst. Graduate Res. Colloquium **2011**, 37–42 (2011). https://doi.org/10.1109/ICSGRC.2011.5991826
9. Segal, M.R.: Machine learning benchmarks and random forest regression (2004)

10. Qi, Y.: Random forest for bioinformatics. In: Zhang ,C., Ma, Y. (eds.) Ensemble Machine Learning. Springer, Boston (2012). https://doi.org/10.1007/978-1-4419-9326-7_11
11. Reis, I., et al.: Probabilistic random forest: a machine learning algorithm for noisy data sets. Astron. J. **157**, 16 (2019). https://arxiv.org/abs/1811.05994
12. Berrar, D.: Bayes' theorem and naive Bayes classifier. Encyclopedia of Bioinformatics and Computational Biology: ABC of Bioinformatics. Elsevier Science Publisher, Amsterdam, pp. 403–412 (2018)
13. Lewis, D.D., et al.: Naive (Bayes) at forty: the independence assumption in information retrieval. ECML (1998)
14. Nugrahaeni, R.A., Mutijarsa, K.: Comparative analysis of machine learning KNN, SVM, and random forests algorithm for facial expression classification. In: 2016 International Seminar on Application for Technology of Information and Communication (ISemantic), pp. 163–168 (2016). https://doi.org/10.1109/ISEMANTIC.2016.7873831
15. Lee, T.R., Wood, W.T., Phrampus, B.J.: A machine learning (kNN) approach to predicting global seafloor total organic carbon. Glob. Biogeochem. Cycles **33**(1), 37–46 (2019)
16. Biswas, A., Chandrakasan, A.P.: Conv-RAM: an energy-efficient SRAM with embedded convolution computation for low-power CNN-based machine learning applications. In: 2018 IEEE International Solid—State Circuits Conference—(ISSCC), pp. 488–490 (2018). https://doi.org/10.1109/ISSCC.2018.8310397
17. Qian, Y., Fan, Y., Hu, W., Soong, F.K.: On the training aspects of deep neural network (DNN) for parametric TTS synthesis. In: 2014 IEEE International Conference on Acoustics, Speech and Signal Processing (ICASSP), pp. 3829–3833 (2014). https://doi.org/10.1109/ICASSP.2014.6854318
18. Snyder, D., Garcia-Romero, D., Sell, G., Povey, D., Khudanpur, S.: X-vectors: robust DNN embeddings for speaker recognition. In: 2018 IEEE International Conference on Acoustics, Speech and Signal Processing (ICASSP) pp. 5329–5333 (2018). https://doi.org/10.1109/ICASSP.2018.8461375
19. Strom, N: Scalable distributed DNN training using commodity GPU cloud computing. In: Sixteenth Annual Conference of the International Speech Communication Association (2015)
20. Pan, J., et al.: Investigation of deep neural networks (DNN) for large vocabulary continuous speech recognition: why DNN surpasses GMMs in acoustic modelling. In: 2012 8th International Symposium on Chinese Spoken Language Processing. IEEE (2012)
21. Voigt, S.: Liquidity and Price Informativeness in Blockchain-Based Markets. Working Paper (2020)
22. Elngar, A.A.: IoT-based efficient tamper detection mechanism for healthcare application. Int. J. Netw. Secur. **20**(3), 489–495 (2018). https://doi.org/10.6633/IJNS.201805.20(3).11
23. Ogu, R.E., Chukwudebe, G.A.: Development of a cost-effective electricity theft detection and prevention system based on IoT technology. In: 2017 IEEE 3rd International Conference on Electro-Technology for National Development (NIGERCON), pp. 756–760 (2017). https://doi.org/10.1109/NIGERCON.2017.8281943
24. Pandit, S., et al.: Smart energy meter using Internet of Things (IoT). Vishwakarma J. Eng. Res. **1**(2), 125–133. Retrieved from http://103.97.164.116:10028/index.php/vjer/article/view/24
25. Bagci, I.E., et al.: Using channel state information for tamper detection in the Internet of Things. In: Proceedings of the 31st Annual Computer Security Applications Conference (ACSAC 2015). Association for Computing Machinery, New York, pp. 131–140 (2015). https://doi.org/10.1145/2818000.2818028
26. Zheng, Y., Dhabu, S.S., Chang, C.-H.: Securing IoT monitoring device using PUF and physical layer authentication. In: 2018 IEEE International Symposium on Circuits and Systems (ISCAS), pp. 1–5 (2018). https://doi.org/10.1109/ISCAS.2018.8351844
27. Kim, S.R., Kim, J.N., Kim, S.T., et al.: Anti-reversible dynamic tamper detection scheme using distributed image steganography for IoT applications. J. Supercomput. **74**, 4261–4280 (2018). https://doi.org/10.1007/s11227-016-1848-y
28. Breitenbacher, D., et al.: HADES-IoT: a practical host-based anomaly detection system for IoT devices. In: Proceedings of the 2019 ACM Asia Conference on Computer and Communications

Security (Asia CCS'19). Association for Computing Machinery, New York, pp. 479–484. https://doi.org/10.1145/3321705.3329847
29. Singh, V., et al.: IoT-Q-band: a low cost internet of things based wearable band to detect and track absconding COVID-19 quarantine subjects. EAI Endorsed Trans. Internet of Things **6**(21), 4. ISSN 2414-1399
30. Hussan, M., Parah, S.A., Gull, S., et al.: Tamper detection and self-recovery of medical imagery for smart health. Arab. J. Sci. Eng. **46**, 3465–3481 (2021). https://doi.org/10.1007/s13369-020-05135-9
31. Kamatagi, A.P., Umadi, R.B., Sujith, V.: Development of energy meter monitoring system (EMMS) for data acquisition and tampering detection using IoT. In: 2020 IEEE International Conference on Electronics, Computing and Communication Technologies (CONECCT), 2020, pp. 1–6 (2020). https://doi.org/10.1109/CONECCT50063.2020.9198495
32. Wang, W., Lu, Y.: Analysis of the mean absolute error (MAE) and the root mean square error (RMSE) in assessing rounding model. In: 2018 Conference Series: Materials Science And Engineering, vol. 324, p. 012049
33. Qi, J., Du, J., Siniscalchi, S.M., Ma, X., Lee, C.-H.: On mean absolute error for deep neural network based vector-to-vector regression. IEEE Signal Process. Lett. **27**, 1485–1489 (2020). https://doi.org/10.1109/LSP.2020.3016837
34. Blanchet, J., et al.: Multivariate distributionally robust convex regression under absolute error loss. Adv. Neural Inf. Process. Syst. **32,** 11817–11826 (2019)
35. Mazzei, D., et al.: A Blockchain Tokenizer for Industrial IOT trustless applications. Future Gener. Comput. Syst. **105**, 432–445 (2020). ISSN 0167-739X. https://doi.org/10.1016/j.future.2019.12.020
36. Lim, J., Kim, Y., Yoo, C.: Chain veri: blockchain-based firmware verification system for IoT environment. In: 2018 IEEE International Conference on Internet of Things (iThings) and IEEE Green Computing and Communications (GreenCom) and IEEE Cyber, Physical and Social Computing (CPSCom) and IEEE Smart Data (SmartData), pp. 1050–1056 (2018). https://doi.org/10.1109/Cybermatics_2018.2018.00194
37. Kaur, E.: IOT Sensing data for recycled water Version 1 (2021 June), from https://www.kaggle.com/ekleenkaur17/iot-sensing-data-for-recycled-water
38. Chaurasia, A.K., Pandey, H.K., Tiwari, S.K., et al.: Groundwater quality assessment using water quality index (WQI) in parts of Varanasi District, Uttar Pradesh, India. J. Geol. Soc. India **92**, 76–82 (2018). https://doi.org/10.1007/s12594-018-0955-1
39. Hong, Z., Chu, C., Zhang, L.L., Yu, Y.: Optimizing an emission trading scheme for local governments: a Stackelberg game mode and hybrid algorithm (2017)
40. Asikgil, B., Erar, A.: Regression error characteristic curves based on the choice of best estimation method. Selcuk J. Appl. Math. (2013)
41. Hastie, T., Tibshirani, R., Friedman, J.: The Elements of Statistical Learning: Data Mining, Inference, and Prediction. Springer, NY (2001)
42. Breiman, L., Friedman, J., Ohlsen, R., Stone, C.: Classification and Regression Trees. Wadsworth, Monterey (1984)

Chapter 4
The Power of Augmented Reality for Smart Environments: An Explorative Analysis of the Business Process Management

Maria Cristina Pietronudo and Daniele Leone

Abstract The chapter proposes an explorative analysis of the use of augmented reality for smart environments highlighting the twofold capability of augmented reality in improving business processes and business environments. Thus, the research question is: "how does augmented reality forge smart environments by improving the business process?" We report evidence from various fields of application such as military, medicine, architecture, automotive, and retail. Indeed, the results show improvements and influences of augmented reality in terms of information management, planning and control, change process, knowledge and performance management, people and customer management, providing a cross-analysis between four dimensions of an organization: resources, process, people, and customer. These improvements enhance business workplaces and connected activities such as learning rooms, online and offline stores, operational rooms, and manufacturing and maintenance departments. The evidence from this study suggests that companies should analyze changes brought by augmented reality about the business process and redesign process to ensure a beneficial impact on the whole business.

Keywords Augmented reality · Virtual reality · Business processes · Technology · Smart environments

4.1 Introduction

Augmented reality (AR) comprehends a variation of virtual reality (VR), where the users are entirely immersed inside the virtual environment [1]. AR has been defined [2] as a system that presents at least three characteristics: the combination of

M. C. Pietronudo · D. Leone (✉)
Department of Management and Quantitative Studies, University of Naples Parthenope, Naples, Italy
e-mail: daniele.leone@uniparthenope.it

M. C. Pietronudo
e-mail: mariacristina.pietronudo@uniparthenope.it

© The Author(s), under exclusive license to Springer Nature Switzerland AG 2022

G. Marques et al. (eds.), *Machine Learning for Smart Environments/Cities*, Intelligent Systems Reference Library 121,
https://doi.org/10.1007/978-3-030-97516-6_4

real and virtual, interactivity in real-time, and the presence of three-dimension. AR can be regarded as a continuum relating purely virtual environments to purely real environments [3].

The research literature on augmented reality is growing [4–6]. Some management scholars have deepened their studies by analyzing augmented reality's role in enhancing museum experiences and purchase intentions [7]; others described how augmented reality creates a more meaningful consumer-brand relationship [8]. Existing literature has, in fact, discussed how AR changes environments, focusing on marketing aspects, e.g., consumer relationships, and the consumer experience. This conceptual chapter investigates how AR affects the business process and transforms the business environment into smart environments. More precisely, this chapter aims to show the AR smart elements that transform the business environment into smart environments through business process improvements (BPIs).

As affirmed by Harrington et al. [9] BPI refers to "a methodology that is designed to bring about step-function improvements in administrative and support processes using approaches such as process benchmarking, process redesign and process re-engineering". The improvement of the process can refer, in practice, to analyze and redefining the sequence of phases forming the previous process.

While a smart environment is intended as a knowledge-based environment that develops extra-ordinary capabilities to be self-aware [10]. It anticipates the actions of a human inhabitant and then automates them [11] using information and communication technology (ICT) and the internet of things (IOT).

Indeed, we focus on the impact of AR on managerial aspects rather than marketing aspects. We contribute in fact to (i) the literature on AR, enlarging the AR spectrum in the business environment; (ii) the literature on BPIs showing the business process vulnerability to certain technology adoption, i.e., AR; (iii) the literature on smart environments offering an overview of the role of AR in business environments. Thus, the research question of this chapter is: "how does augmented reality forge smart environments by improving the business process?"

To tackle this research question, we review the literature on augmented reality and trace the historical background from the 60s to the present to analyze the main differences and applications between virtual reality and augmented reality that differently influence environments and business units. Secondly, as already mentioned, the chapter pays attention to the BPIs using AR and understanding how technology can help and facilitate the work of managers in the search for competitive advantage and where (in which workspace) AR operates.

The results show improvements and influences of AR in terms of Information Management, Planning and Control, Change Process, Knowledge Management, Performance management, People and Customer Management; providing a cross-analysis between four dimensions of an organization: resources, process, people and customer. The influence of AR on the organizational dimension and the business unit transforms the certain business environment into smart environments such as learning rooms, online and offline stores, operational rooms and manufacturing and maintenance departments.

Managerial implications are directed to the AR users of international firms and suggest registration and sensing errors as two of the most significant limitations in building effective augmented reality systems. This chapter summarizes current efforts to overcome these problems. Thus, we focus on analyzing the business process improvements, exploring various fields of applications such as the military, medical, entertainment, and automotive industries that have been explored in several international firms.

This work is structured into the following parts. The next section presents the research approach and Sect. 3 reviews the theoretical background and first evidence from the literature review. Section 4 and Sect. 5 present evidence from history about the use of augmented reality for business process management and evidence from practical applications of AR in various sectors and business units. The Sect. 6 closes the chapter with implications and conclusions of the study.

4.1.1 Research Design

This chapter aims to uncover the improvements developed by augmented reality on business processes management and consequently on the business environments. The research design is based on the analysis of management theories, historical background and case studies. After that, we aim to provide a conceptual framework synthesizing business processes components and respective improvements observed by the application of AR in the empirical evidence discussed in the following sections. Indeed, as described by Gilson and Goldberg [12] "*beyond summarizing recent research, manuscripts should provide an integration of literatures, offer an integrated framework, provide value added, and highlight directions for future inquiry. Papers are not expected to offer empirical data.*"

Drawing on these assumptions, we traced the history of the augmented reality origins is performed from 1962 (the first experience of AR) to the present day. After that, multiple case studies in various sectors and business units demonstrate an exploration of the main fields of usage of AR.

4.2 The Historical Evolution of Augmented Reality for Smart Environments

AR is a technology that originated many decades ago; its evolution has gone through various stages before reaching today, going from science fiction to real and effective innovation (Fig. 4.1). For its nature, AR transforms physical environments into augmented environments in which virtual and real dimensions are interrelated. One thinks of AR in shops, museums, cinemas and at how it transforms the consumer experience. However, the technology has been improved by complementary technologies

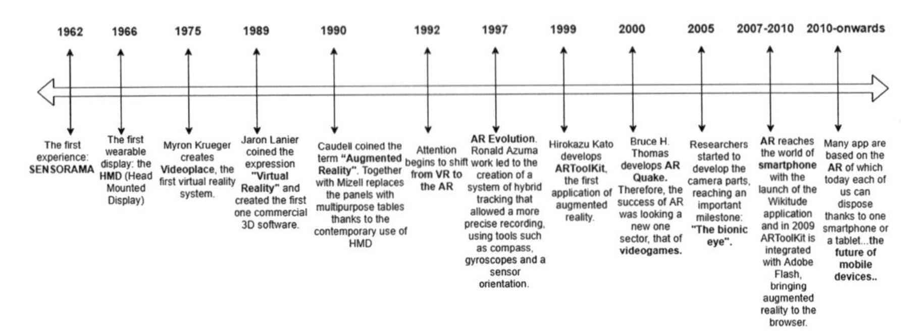

Fig. 4.1 The AR evolution. The figure shows the AR evolution, marking the shift from VR to AR and from science fiction to real and effective innovation

such as artificial intelligence that has empowered AR output, shifting augmented environments into a smart environment able to process data and information rather than merely visualize data. We retrace the most important phases of this technological evolution, showing how integration with intelligent technology has empowered AR to transform physical environments into ever more smart environments.

AR is an incremental evolution of Virtual Reality (VR) born to enhance the entertainment industry. In 1962, a film director named Morton Helig invents Sensorama, a machine able to transform the visual experience into immersive. The machine deployed 3D images, stereo sounds, tactile and olfactory sensations, allowing people no longer to be passive in watching a movie but to interact with it and use all the senses available, just as if they could perform determined actions. In 1966, Professor Ivan Sutherland (Harvard University) invented the first wearable display: the HMD (Head Mounted Display). A kind of helmet allows the user to immerse himself in images projected on the screen through a monocular or binocular optic device. In 1975, Myron Krueger creates Videoplace, the first virtual reality system. The movements of people are recorded by a video camera, analyzed and transmitted to the virtual environment built by the computer; from this, it follows correspondence between movements and representations on the screen that transmits in the minds of users the feeling of interaction with virtual objects. However, the inventions mentioned above were still immature, not fully ready for the market. Instead, in 1989 Jaron Lanier coined the expression "Virtual Reality" and created the first commercial 3D software. It was the first commercial case in which someone managed to monetize virtual reality, thus beginning a new business in virtual worlds. These VR solutions were changing the entertainment environment, transforming it into a parallel environment that was neither interactive nor smart. In 1992, the attention become to shift from virtual reality to augmented reality. The user is no longer immersed in an environment created at the computer, but the information comes out of the virtual world and invades reality. In this manner, tech devices allow you to add information layers to the physical world, transforming it into a more interactive place. However, the first advancements of AR concerned the users' movements in a mix of physical and digital places. In 1997, Professor Ronald Azuma publishes a survey on augmented reality exploring key characteristics and problems. The article is still one of the most 50 influential articles in the history of the MIT press. It gave the ride for a broad discussion on the future of this powerful technology since it identified challenges that needed to be solved before AR fulfilled its ultimate potential. In other terms, Azuma's academic contributions contribute to receiving AR as a technology able to improve the user's understanding of and interaction with the real world [13]. In the following years, first applications of AR were diffused. In 1999, Hirokazu Kato develops ARToolKit, an application that, using video-tracking, can be positioned with a virtual camera in the same point of view as the observer and on this add information superimposed. In 2000, Bruce H. Thomas develops AR Quake [14], a version of the Quake game based on AR technology. A wearable device enables the user can move into the real world by superimposing the subjective images of the videogame. Therefore, the success of augmented reality was looking like a new sector, that of video games. AR, in fact, was changing entertainment environments until 2005 when the medical field began to perceive the usefulness of technology placed at the service of man. A team of scientists reached an important milestone with "The bionic eye" [15]. It was a retinal prosthesis system for stimulation of the retina with a resolution corresponding to a visual acuity of 20/80-sharp enough to orientate towards objects, recognize faces, read large fonts, watch TV and, perhaps most importantly, lead an independent life. In this sense, AR connected better users to the people, locations and objects around them, rather than cutting them off from the surrounding environment.

In 2008, augmented reality also reached the smartphone world with the launch of the Wikitude application. This software allows you to view information on your mobile phone published by other users on Wikipedia and Panoramio, referring to the site in which they are located. In 2009, ARToolKit is integrated with Adobe Flash, bringing augmented reality to the browser. From 2010 onwards, the AR systems have made major changes concurrent with the whole technology sector, that of smartphones and social networks. From many companies in the industry, augmented reality is seen as the future of mobile devices. In accordance with Azuma [13], AR is the most likely route by which AR displays might replace all other display form factors: monitors, laptop screens, phones and tablets, becoming a pervasive technology that transforms different environments such as homes, workplaces, entertainment spaces, cities and so on.

Thus, while VR technology can realize imaginary environments or reconstruct structures or parts of them that no longer exist [16], AR systems increase the user's perception and interaction by providing additional information that he could not obtain simply and directly using his senses [17].

This historical background confirms that a virtual 3D reconstruction allows bringing an object back to life a structure with the characteristics it had at its origins and to be able to see its changes or deterioration in the slightest detail [18]. In fact, from its first use up to the present day, VR has been widely used in many professions. For instance, archaeologists and historians who, thanks to the knowledge and the help of this technology, have allowed everyone to view the actual dimensions and features of some dating structures and cities at different historical periods. Therefore, in modern architecture, VR has allowed architects to realize their projects in a more detailed way and present them through a computer in 3D mode [19], a technique that cannot be reproduced using traditional paper and pen. Consequently, VR enhances the comprehension of the environment; on the other side, AR enhances the environment's capabilities in interacting with other environment components such as humans, objects, devices and places. In other terms, AR transforms the environment into a smart place that interacts with exchanging information and data.

As a result, the real world is "virtually" enriched with information generated and synchronized by the computer about the real object to which the vision device is directed.

4.3 Augmented Reality and Business Process Management

The adoption of new technology very often requires a contiguous transformation of the organizational and business processes. This transformation is essential to trigger process improvement or improve business results [20]. The themes of transformation and improvement of business processes—comprehensively examined in management studies—are by their nature connected to those of technologies and innovations. Business process management since the 90s is, in fact, proposed as a discipline that supports and improves business processes employing principles, methods and techniques through which it designs, implements, controls and analyzes operational processes [21–23]. Specifically, it combines IT, management science and industrial engineering so that a transformation includes all the resources participating in the business, from employees to partners, from IT systems to machinery [24].

Many scholars have evaluated the impact of technological innovations on the business process. Already Davenport [25] considered technology as an enabler of process innovation. Subsequently, other authors stressed the importance of blending technological innovations with process innovations to benefit the whole organization [26]. The most studied field was that of IT technology, which generated a series of changes in the way of transferring information and knowledge and facilitating organizational learning as a whole [27]. Scholars attested to a very close relationship between technologies and business processes. If, on the one hand, the business process management has to adapt to new technologies, the technologies must be well integrated with the business processes. VR and AR are also valuable tools to improve and accelerate products and development processes in many industrial applications [28]. Recently, consumer-grade immersive VR and AR technologies have become more accessible and cost-effective, creating mass industrial interest in these technologies [29]

Precisely, ARs can imply an improvement in interaction, involvement [30], and coordination between the organizational units and between partners. However, studies that explicitly evaluate the impact of AR on the business process are very few. Bogdan and Popovici's study [31] pointed out the importance of describing the entire business process, the formulation of the business rules, the business object, and the identification of stakeholders who interact with the system before implementing an augmented reality technology. However, the study is premature and very specific of the entire process design of Virdent, software to increase the quality of the educational process in dental faculties.

Recent studies are discussing major challenges for the implementation of AR [32, 33]. Samini et al. [33] had linked business efficiency and AR, showing performance metrics to measure the efficiency of processes improved by AR (task completion time, learnability, user comfort, difficulty to operate, accuracy, the number of operational errors or misalignment or the number of times a user skipped a task). Hadar [34] observed development methods for AR, yet did not include streamlining business process steps but were focused on constructing machine-readable instructions for presenting AR digital overlay and communicating with devices by AR content authoring and interpretation tools. As a result, he raised the issue of setting functional and non-functional requirements for reimagining business processes and driving the business's competitiveness.

A first attempt to analyze the improvement of the business process due to AR is given by Zhao et al. [35]. The authors illustrate a model based on better business processes and human resources management using VR and AR tools. Model is based on the synergistic effect of the use of VR/AR simultaneously in human resources processes (recruitment, development, training, staff evaluation) and the main operational processes: production logistics, sales, etc., which will ultimately reduce production costs, develop employees and increase the cost of the organization's intellectual (which, for example, can be expressed in the rising cost of services provided by the company).

4.3.1 Smart Environment

Smart environment is intended as a knowledge-based environment that develops extra-ordinary capabilities to be self-aware [10]. Furthermore, it is an environment that anticipates the actions of a human inhabitant, automating their actions [11] and integrating them better with the augmented space. The concept is often overlapped with those of the smart city. However, groups of scholars agree in considering a smart environment as a transversal concept involving homes, cities, and factories [36].

The common point is the use of ICT and IoT technology to facilitate communication among different entities. Kumar [10] identifies technological elements that convert an environment into a smart one:

- remote control of devices and a power line communications systems;
- device communication to connect environmental elements;
- information acquisition and dissemination from sensors;
- enhanced services by intelligent devices;
- predictive and decision-making capabilities.

Most recent studies [37] consider the potentiality of AR in smart environments, emphasizing its capability in creating novel and enhanced IoT interfaces that may offer greater accessibility over traditional interaction techniques [38, 39]. AR, in fact, is considered as an integrative technology that improves mainly IOT interfaces. Authors [37] state AR in smart environments can:

- visualize relevant information regarding the state of IoT devices in the user's environment;
- provide additional means of embodied input for users to interact with these devices, such as via sunglasses or hand tracking.

4.4 Case Studies from Different Sectors

4.4.1 Methodology

The study is a collective case study where several cases were chosen to investigate the topic under investigation [40]. The collective case study has been used as a methodology by many recent empirical works investigating innovation processes or innovation factors [41, 42]. The research has an exploratory nature.

We intend to explore the potentiality of AR in transforming the business environment into smart environments through BPIs. Mainly we address the following research question: "how does augmented reality forge smart environments by improving the business process?"

Our empirical setting followed conceptual considerations [43] also linked to the degree of appropriateness with our research question [44]. To define our analysis sample, we adopt a purposive [43] rather than a statistical sampling logic [45], adopting a non-probability sampling technique. We select height cases study focused on the fields where AR technology has represented a real innovation for improving business models of different organizations [46]. Further criteria are applied to select case studies that better explain the contribution of AR in business environments: we choose cases that provide good access to secondary information and most indicative cases to explain AR and BPIs. Although our analysis is based on secondary data, secondary data have become easier to collect and more accurate and more available to perform analyses of all kinds [47]. However, data were triangulated [48] from multiple sources of evidence i.e., observations, online documentation and online video recording.

4.4.2 Cases Study Description

We explore AR applications in business environments, particularly workplaces.

Two of the last major innovations of AR known in the military field, turn out to be iOptik and Red 6 Aerospace. iOptik is a project of the Innovega company, specializing in creating augmented reality viewers, funded by DARPA (Defense Advanced Research Projects Agency). This system is equipped with a pair of glasses structured in polycarbonate with two contact lenses with special filters. Their polycarbonate structure gives greater strength and lightness, ensuring that comfort of use; through the lenses, images in augmented reality are projected, with the ability to adjust the action of the filters to allow the passage of various types of light. Thus, these special glasses allow the military to see digital information and the real environment around them simultaneously. Furthermore, a filtering level can be set to cancel the view of the real environment and immerse the user in reality only digital. This technology provides key information in real-time, codifying (in real-time) essential data connecting better the users to the external environment.

The Red 6 Aerospace software simulates flying in an open system (the sky), overcoming the traditional flight simulators, which train pilots when stationary. Aviators learn quickly through more responsive digital methods and require fewer traditional resources such as instructors and certain equipment. Instead of connecting users to a closed, internal system, the software allows pilots to go and experience flight physically and simulate real situations. In contrast, their trainers are connected in augmented reality. The software creator claims that "baby drivers who go through their training program, with huge cost savings for training providers and with huge efficiencies", as well as avoiding the dependence of the armed forces on expensive simulators. The Red 6 Aerospace AR technology integrates the physical environment with digital elements, enriching training places such as traditional classrooms.

The field of medicine is a sector that is always alive in terms of innovative technology, developed to assist doctors and surgeons in their work and in patient care. The field of surgery, for example, in many operations requires internal images that are transmitted by a small camera. And even augmented reality, in recent years, is looking to expand in this field to facilitate and optimize surgeries by surgeons.

Google Glass is something different than the direct video systems already used by everyone in hospitals, indeed, several improvements arise from the use of the technology: firstly, the people's engagement and involvement in the surgery or the training process enhance the medical training; secondly, the information available in real-time for surgeons and their staff consequently speeds up the transfer and the sharing of information and tacit knowledge. This last point may expedite a typical pre-operative phase delivering effective information to surgeons and nurses; to put it plainly, it reduces the time a surgeon spends to collect information on the patient. In other terms, AR transforms operating rooms into smart operating rooms and augmented teaching rooms at once.

Another sector always looking for technological innovations is that of architecture and design. For many years computers have existed that allow us to work on the structure and presentation of architectural and architectural works building constructions, also in 3D. Still, through the technique of previously exposed markers, the AR systems have given the possibility to the big ones building companies and architects to show their projects anywhere you want and to anyone. This innovation has improved relations with customers since showing a 3D model, rotatable through 360°, allows one better space planning and design visualization, important in choosing a project.

The combination of augmented reality—design has been a gamble on which it has wanted to also aim for IKEA, the leader in the design industry. The goal of the Swedish company has always been to try to please its customers, providing them with the best services and functions that will help them in their choice. Recently, IKEA has tried to respond to the problems it comes across more often than not when choosing a product: "*How's it inside?*", "*Can it enter without space problems?*"

The company wanted to solve these dilemmas with augmented reality systems, following a new marketing strategy based on three elements: the paper catalog, the IKEA application and a smartphone (or tablet). In fact, after selecting the piece of furniture, just scanning the code connected to it via the smartphone camera, instantly its internal version will appear on our screen in a 3D version. We can move and rotate to our pleasure in the real environment to which it overlaps, transmitting the same sensations of a real and concrete object tried in a determined space of the house. In doing so, IKEA increases customer involvement with immersive purchase experience, also affecting customer satisfaction. At the same time, IKEA contributes to enhancing the users' environments by transforming users' homes in IKEA shops.

The automotive sector is continuously in close contact with the technology in all production processes. BMW and Volkswagen were among the first automotive companies to invest in augmented reality to improve production processes, plan production lines, and check various car components.

Obviously, for more experienced employees, this could not be difficult, but even replacing a simple component can result in difficulties. Therefore, thanks to special glasses, the mechanic must look at the point of interest. The application provides one series of additional content overlapping with real images shown on display to guide operation. Improvement in assembly, maintenance and repair phases are of various kinds. AR reduces the maintenance lines and human errors, optimizing decision-making processes for the entire post-productive process; indeed, it is useful for employees who learn by doing and processing effective information reducing the repair time. The aforementioned automotive companies represent, in fact, virtuous examples of smart factories that, with AR, enhance the working men's environment.

Some large distribution companies saw AR improve and make the experience more interactive and enjoyable purchase for every consumer. The companies provide this experience through a special packaging of their products. They are recognized by AR devices instantly, with the possibility to help the buyer in the choice of purchase or to provide interactivity and fun with the product afterward have bought it. Concerning the use of AR to provide additional information, it is possible to mention the LEGO case. The Danish company, a precursor of smart packaging, wanted to provide further safety when choosing their toys. Despite a graphic representation of the toy on each box, we do not always realize how the version will be final and concrete of the toy purchased. To solve this dilemma, LEGO has installed a screen inside its stores that can recognize the boxes that appear before them and reproduce the toy in a 360° rotatable 3D model, thus giving the consumer a future digital version of the toy. This new marketing strategy, initially experimental, proved to be successful since these systems of AR attracted every consumer before reaching the cash desk.

From this perspective, the Tesco case study and others can also be connected based on LEGO's same basic concept, i.e., supply additional information on the product you intend to purchase, framing it with a smartphone or a tablet. For example, Tesco has applied IBM's augmented reality on its shelves by providing content with additional products framed. Referring to the second type of smart packaging, the aim is to go beyond the shopping experience, linking brand and consumer even after using the product. Examples of such marketing strategies are so many now. Mr. Heinz sells its products with a special packaging which provides recipes and tips on the sauce you intend to buy; RedBull, putting inline its cans, gives the possibility to create its circuit and then play it like in a racing machine videogame. Some people have seen smart packaging as a way to participate in prize competitions.

The innovations of augmented reality dedicated to the final consumer are in the retail sector, downsizing the shopping experience, focusing on the doubts that arise in choosing one product. However, AR solutions may improve customer involvement and brand interaction by offering immersive purchase experiences in their physical shops. They furthermore provide additional information to the consumer during the purchase stage, increasing their awareness of the product (i.e., Tesco experience) or brand (i.e., Lego). To conclude, AR may reinforce the effectiveness of marketing strategies.

4.5 Discussion

As far as the methodological choice is concerned, Table 4.1 shows remarkably interesting evidence. Indeed, it synthesizes business process components [49] and respective improvements observed due to the AR application in business management and the environment. The table shows AR capabilities that convert a business environment into a smart environment relative to the environment. Furthermore, the table clarifies that an improvement in a specific business component may affect a different business component, simultaneously enhancing the business environment as a whole.

The information available in real-time transpired in medicine and military sectors and the data codification and the integration of data and information that flow in the virtual and real environment. These imply improvements in information management: a new set of data and info are processed, stored, and made available to transfer, when needed, information but also tacit knowledge to AR users. In many cases, as in automotive, improvements in information management may optimize the decision-making process of workers at the operational level. The different actors involved in business processes have always used tools in common without a technology that supports the actual smart environments. They have to put their knowledge and skills

Table 4.1 Business process improvements relative business process components and environment

Business process		Applications of Augmented reality		Environment	
Business process component	Improvements from AR	Case study	Sectors	Application environment	Smart elements
Information management	Data codification in real time	iOptik	Military	Military arena, operational rooms, teaching rooms, manufacturing and maintenance department	Data codification
	Information availability in real-time	iOptik	Military		Real-time visualization
		Google glass	Medicine		
	Integration of data and information in real-time	Red 6 aerospace	Military		Real-time input provider
	Optimizing decision-making processes	BMW—Volkswagen	Automotive		Real-time output provider
	Transferring and sharing information and tacit knowledge in real-time	Google glass	Medicine		Real-time input/output provider
Planning and control	Optimizing the maintenance lines	BMW—Volkswagen	Automotive	Manufacturing and maintenance department	Real-time input/output provider
	Streamlining the control process		Automotive		Decision-making capabilities
Change process	Expediting the pre-operative phase with effective information	Google glass	Medicine	Operational rooms, manufacturing and maintenance department	Real-time input/output provider

(continued)

Table 4.1 (continued)

Business process		Applications of Augmented reality		Environment	
Business process component	Improvements from AR	Case study	Sectors	Application environment	Smart elements
	Expediting the production control phase	BMW—Volkswagen	Automotive		
Knowledge management	Reducing the gap between training and experience	Red 6 aerospace	Military	Military area, teaching rooms, manufacturing and maintenance department	Real-time visualization
		BMW—Volkswagen	Automotive		
	Improving the training process	Google glass	Medicine		Real-time input/output provider
	Effective information	BMW—Volkswagen	Automotive		
Performance management	Saving training costs improving pilot performance	Red 6 aerospace	Military	Teaching rooms, manufacturing, and maintenance department, Store	Adaptation to environment behavior
	Reducing the time of repair	BMW—Volkswagen	Automotive		
	Optimizing post-productive processes				
	Effectiveness of marketing strategies: increasing product or brand awareness	Lego, Tesco	Retail		

(continued)

Table 4.1 (continued)

Business process		Applications of Augmented reality		Environment	
Business process component	Improvements from AR	Case study	Sectors	Application environment	Smart elements
	Effectiveness of marketing strategies: improving consumer-brand interaction	LEGO, Heinz, RedBull	Retail		
People	Reducing human errors	BMW—Volkswagen	Automotive	Manufacturing and maintenance department, military arena, teaching rooms	Decision-making capabilities
	People engagement and involvement	iOptik	Military		Real-time visualization
		Google glass	Medicine		
Customer management	Customer satisfaction	IKEA	Architecture and design	Customers' home, store, maintenance department	Entity virtualization
	Customer involvement and immersive purchase experience		Architecture and design		
		LEGO, Heinz, RedBull	Retail		
	Customer experience	BMW—Volkswagen	Automotive		

The table shows business process improvement due to AR and AR capabilities that convert certain business environment into a smart environment at the service of the community by using smart IT solutions as AR and handing down stories of business process improvements to the other firms' actors. In terms of environment, the AR transforms the military arena, operational rooms, teaching rooms, and manufacturing and maintenance departments into smart environments, codifying external data and visualizing them in real-time. In addition, the integration of AR with other advanced technology facilitates the communication input and output data of integrated systems in real-time.

Improvements have also been observed in planning and controls, particularly in the automotive sector. AR optimizes the production/maintenance lines digitalizing the existing production process in manufacturing firms. It streamlines the control process introducing real-time control rather than post-release control. Evidence shows that AR in planning and control operates mainly in the manufacturing and maintenance environment. AR manifests smart elements that transform traditional environments into smart environments: its ability to provide input and output to other systems in real-time and the decision-making for operational tasks.

In terms of the change process—changing the organization to meet users' requirements—AR expedites several phases of the business process. AR technologies speed up the pre-operative phases in medicine, which usually require a time slot to collect and verify patients' clinical information, producing effective information for nurseries and surgeons in real-time. In automotive, the production control phases ensure that the process's execution is done optimally during the action phase. Therefore, AR makes smart operational rooms and manufacturing and maintenance departments (again) connecting humans, devices, and advanced technologies present in that specific workplace.

In other circumstances, AR improves knowledge management, reducing the gap between training and experience, promoting techniques of learning as experiential learning. Users acquire knowledge and competencies combining intellectual comprehension and practical experience. Furthermore, AR helps users to make effective use of information and data. AR enhances teaching rooms and maintenance and manufacturing environments, making them a workplace where learning by doing practices are ordinary practices. Smart elements concern the capability of AR in visualizing data in real-time to monitor task performance and capture feedback from other connected systems in real-time.

Regarding performance management, augmented technologies reduce costs allocated for training and time of work, increasing single-user productivity and the entire organization. AR impacts the effectiveness of marketing strategies increasing consumer-brand interaction and product or brand awareness in the retail setting. In the first case, environment enhancements concern teaching rooms and manufacturing and maintenance environments; in the second case, further business environments can be positively affected by AR, i.e., stores. In terms of performance management, AR appears to adapt its output to the environmental behaviors transforming the environment into a smart one.

With the reference of people, AR tools guide workers in each phase of their task, reducing human errors enabling workers to accomplish tasks with less pressure and increasing engagement and involvement during the training phase and the working hours. These improvements concern environments such as the manufacturing and maintenance department, military arena, teaching rooms where AR supports people in a twofold manner: visualizing information in real-time and receiving informed support in decision making.

Finally, improvements in customer management are more deducible and concern architecture and design or retail (specifically packaging operational strategy). AR creates customer involvement and immersive purchase experience, providing interactive and additional information about products or reproducing occasions of use in an augmented world. Therefore, customer satisfaction may increase when immersive technologies support the customer journey. As shown by marketing studies, AR forges new concepts of stores, sometimes leveraging on multi-channel retailing. For instance, in the case of IKEA, it transforms the consumers' homes into an interactive shop. The main element that leads AR to create a smart environment is its capability to reproduce and visualize virtual entities in physical spaces.

The uses of augmented reality are not limited to a single goal or sector. This technology directly influences company business processes, from producing a new product, a new marketing method to a new approach with consumers. Therefore, augmented reality can affect innovation but can mainly lead an organization to innovate its business processes [46] and simultaneously transform the traditional business environment into a smart one. Since the 1990s, augmented reality has gone from a simple scientific laboratory experiment for computer enthusiasts to real innovation in many fields. Based on the synchronization between the real world and digital, these technological systems have not changed how a company operates. They have undoubtedly represented an innovation that has improved, and in some cases facilitated, the work of particular business processes, receiving later estimated approvals for the contribution made [7, 8]. Instead, we enlarge the AR spectrum in the business environment as a whole.

Other than showing the difference between virtual and augmented reality and supporting the limitation of VR in the use of certain functions, the chapter provides an analysis of improvements produced by AR in the business process of some sectors and the business environment. Augmented reality is not limited to just some specific uses; on the contrary, in the last decade, AR applications have always been more advanced and extended to various fields of application and devices that display integrating, virtual and real environments. These integrations cause a flow of data that changes the way to process and diffuse knowledge in an organization, the way to define business rules or business activity. The implementation of an approach based on business processes to interpret the adoption of AR technology illustrates a series of advantages as the introduction of experiential learning, the reduction on the production line, or the saving of costs and time of production, making apparent an impact of AR on several business process components.

AR effects are analyzed across all organization dimensions [49]: resources, processes, people, and customers. The resource view focused on data and information used in an organization and processed or available (in real-time) by AR tools. The process view includes the way we plan, control, and change business processes after AR. The people's view includes changes in terms of involvement, engagement, and worker productivity. The customer view deals with improvements in the purchase experience.

4.6 Conclusion

Starting from the business model improvements induced by AR, we highlight AR's potentiality in transforming the business environments into smart environments. Our explorative analysis focuses on the support of AR in smart environments, especially to implement innovative IoT interfaces able to monitor relevant information of the users' environments in order to create a unique and smart context in a value co-creation perspective. AR' smart elements intended as AR capabilities—such as data modification, real-time visualization, real-time input/output provider, decision-making capabilities, adaptation to the external environment, and entities virtualization -improve the learning process, absorption process, and engagement process acting on the traditional workplace. Evidence shows relevant changes in teaching rooms, operational rooms, manufacturing and maintenance stores, and customers' homes. This chapter is linked to the main topic of this book because, in these environments, AR does not operate only to enhance the user experience but also to transform the environment into a smarter and more interactive place. Taking evidence together, managers should adopt a business process approach in adopting AR. AR device manufacturers should primarily focus on the adaptability and flexibility of AR technologies with the company's needs and business architecture; researchers should deeply explore AR's effect on BPIs. For these reasons, this chapter also emphasizes novel managerial issues, usually far from traditional practices of business process management, which are increasingly becoming relevant for the development of different industries over time. Future research should further explain how AR smart components impact smart environment improvements investigating new industries or connecting these themes to firms' performance.

References

1. Zhou, F., Duh, H.B.L., Billinghurst, M.: Trends in augmented reality tracking, interaction and display: a review of ten years of ISMAR. In: Proceedings of the 7th IEEE/ACM International Symposium on Mixed and Augmented Reality, pp. 193–202. IEEE Computer Society (2008)
2. Azuma, R.T.: A survey of augmented reality. Presence Teleoperators Virtual Environ **6**(4), 355–385 (1997)
3. Milgram, P., Takemura, H., Utsumi, A., & Kishino, F.: Augmented reality: a class of displays on the reality-virtuality continuum. In: Telemanipulator and Telepresence Technologies, vol. 2351, pp. 282–293. International Society for Optics and Photonics (1995)
4. Zahedi, F.M., Walia, N., Jain, H.: Augmented virtual doctor office: theory-based design and assessment. J. Manag. Inf. Syst. **33**(3), 776–808 (2016)
5. Tredinnick, L.: Augmented reality in the business world. Bus. Inf. Rev. **35**(2), 77–80 (2018)

6. Grzegorczyk, T., Sliwinski, R., Kaczmarek, J.: Attractiveness of augmented reality to consumers. Technol. Anal. Strateg. Manag. **31**(11), 1257–1269 (2019)
7. He, Z., Wu, L., Li, X.R.: When art meets tech: the role of augmented reality in enhancing museum experiences and purchase intentions. Tour. Manage. **68**, 127–139 (2018)
8. Scholz, J., Duffy, K.: We ARe at home: how augmented reality reshapes mobile marketing and consumer-brand relationships. J. Retail. Consum. Serv. **44**, 11–23 (2018)
9. Harrington, H.J., Esseling, E.C., Van Nimwegen, H.: Business Process Improvement Documentation, Analysis, Design and Management of Business Process Improvement. McGraw-Hill, New York (1997)
10. Kumar, V.: Smart Environment for Smart Cities. In: Smart Environment for Smart Cities, pp. 1–53. Springer, Singapore (2020)
11. Youngblood, G.M., Heierman, E.O., Holder, L.B., Cook, D.J.: Automation intelligence for the smart environment. In: International Joint Conference on Artificial Intelligence, vol. 19, p. 1513. Lawrence Erlbaum Associates Ltd. (2005)
12. Gilson, L.L., Goldberg, C.B.: Editors' Comment: So, What is a Conceptual Paper? (2015)
13. Azuma, R.T.: Making augmented reality a reality. In: Applied Industrial Optics: Spectroscopy, Imaging and Metrology, pp. JTu1F-1. Optical Society of America (2017)
14. Piekarski, W., Thomas, B.: ARQuake: the outdoor augmented reality gaming system. Commun. ACM **45**(1), 36–38 (2002)
15. Palanker, D., Vankov, A., Huie, P., Baccus, S.: Design of a high-resolution optoelectronic retinal prosthesis. J. Neural Eng. **2**(1), S105 (2005)
16. Burdea, G.C., Coiffet, P.: Virtual Reality Technology. Wiley (2003)
17. Poushneh, A., Vasquez-Parraga, A.Z.: Discernible impact of augmented reality on retail customer's experience, satisfaction and willingness to buy. J. Retail. Consum. Serv. **34**, 229–234 (2017)
18. Javornik, A.: Augmented reality: research agenda for studying the impact of its media characteristics on consumer behaviour. J. Retail. Consum. Serv. **30**, 252–261 (2016)
19. Bruno, F., Bruno, S., De Sensi, G., Luchi, M.L., Mancuso, S., Muzzupappa, M.: From 3D reconstruction to virtual reality: a complete methodology for digital archaeological exhibition. J. Cult. Herit. **11**(1), 42–49 (2010)
20. Berente, N., Lee, J.: How process improvement efforts can drive organisational innovativeness. Technol. Anal. Strateg. Manag. **26**(4), 417–433 (2014)
21. Pyon, C.U., Woo, J.Y., Park, S.C.: Service improvement by business process management using customer complaints in financial service industry. Expert Syst. Appl. **38**(4), 3267–3279 (2011)
22. Weske, M.: Business process management architectures. In: Business Process Management, pp. 333–371. Springer, Berlin, Heidelberg (2012)
23. Dumas, M., La Rosa, M., Mendling, J., Reijers, H.A.: Fundamentals of Business Process Management, vol. 1, p. 2. Springer, Heidelberg (2013)
24. Smith, H., Fingar, P.: Business Process Management: The Third Wave, vol. 1. Meghan-Kiffer Press, Tampa (2003)
25. Davenport, T.H.: Process Innovation: Reengineering Work Through Information Technology. Harvard Business Press (1993)
26. Clark, T.H., Stoddard, D.B.: Interorganizational business process redesign: merging technological and process innovation. J. Manag. Inf. Syst. **13**(2), 9–28 (1996)
27. Srivardhana, T., Pawlowski, S.D.: ERP systems as an enabler of sustained business process innovation: a knowledge-based view. J. Strateg. Inf. Syst. **16**(1), 51–69 (2007)
28. Ma, M., Zheng, H.: Virtual reality and serious games in healthcare. In: Advanced Computational Intelligence Paradigms in Healthcare 6. Virtual Reality in Psychotherapy, Rehabilitation, and Assessment, pp. 169–192. Springer, Berlin, Heidelberg (2011)
29. Moser, T., Hohlagschwandtner, M., Kormann-Hainzl, G., Pölzlbauer, S., Wolfartsberger, J.: Mixed reality applications in industry: challenges and research areas. In: International Conference on Software Quality, pp. 95–105. Springer, Cham (2019)
30. Poppe, E., Brown, R., Johnson, D., Recker, J.: Preliminary evaluation of an augmented reality collaborative process modelling system. In: 2012 International Conference on Cyberworlds, pp. 77–84. IEEE (2012)

31. Bogdan, C.M., Popovici, D.M.: Information system analysis of an e-learning system used for dental restorations simulation. Comput. Methods Progr. Biomed. **107**(3), 357–366 (2012)
32. Hadar, E.: Toward development tools for augmented reality applications–a practitioner perspective. In: Workshop on Enterprise and Organizational Modeling and Simulation, pp. 91–104. Springer, Cham (2018)
33. Samini, A., Palmerius, K.L.: Popular performance metrics for evaluation of interaction in virtual and augmented reality. In: 2017 International Conference on Cyberworlds (CW), pp. 206–209. IEEE (2017)
34. Van Krevelen, D.W.F., Poelman, R.: A survey of augmented reality technologies, applications and limitations. Int. J. Virtual Reality **9**(2), 1 (2010)
35. Zhao, H., Zhao, Q.H., Ślusarcyk, B.: Sustainability and digitalization of corporate management based on augmented/virtual reality tools usage: China and other world IT Companies' experience. Sustainability **11**(17), 4717 (2019)
36. Gomez, C., Chessa, S., Fleury, A., Roussos, G., Preuveneers, D.: Internet of things for enabling smart environments: a technology-centric perspective. J. Ambient Intell. Smart Environ. **11**(1), 23–43 (2019)
37. Flick, C.D., Harris, C.J., Yonkers, N.T., Norouzi, N., Erickson, A., Choudhary, Z., Gotsackker, M., Bruder, G., Welch, G.: Trade-offs in augmented reality user interfaces for controlling a smart environment. In: Symposium on Spatial User Interaction, pp. 1–11 (2021)
38. Rashid, Z., Melià-Seguí, J., Pous, R., Peig, E.: Using augmented reality and internet of things to improve accessibility of people with motor disabilities in the context of smart cities. Futur. Gener. Comput. Syst. **76**, 248–261 (2017)
39. Kim, K., Billinghurst, M., Bruder, G., Duh, H.B.L., Welch, G.F.: Revisiting trends in augmented reality research: a review of the 2nd decade of ISMAR (2008–2017). IEEE Trans. Vis. Comput. Graph. **24**(11), 2947–2962 (2018)
40. Stake, R.E.: Case study: Composition and performance. Bull. Council Res. Music Educ., 31–44 (1994)
41. Dinesh, K.K., Sushil: Strategic innovation factors in startups: results of a cross-case analysis of Indian startups. J. Global Bus. Adv. **12**(3), 449–470 (2019)
42. Hydle, K.M., Billington, M.G.: Entrepreneurial practices of collaboration comprising constellations. Int. J. Entrepreneurial Behav. Res. (2020)
43. Silverman, D.: Doing Qualitative Research: A Practical Handbook. Sage (2013)
44. Mason, J.: Qualitative Researching. Sage (2017)
45. Bryman, A.: Quantity and Quality in Social Research, vol. 18. Routledge (2003)
46. Leone, D., Pietronudo, M.C., Dezi, L.: Improving business models through augmented reality applications. evidence from history, theory and practice. Int. J. Qual. Innov., vol. ahead-of-print No. ahead-of-print (2021). https://doi.org/10.1504/IJQI.2021.10037677
47. Wamba-Taguimdje, S.L., Wamba, S.F., Kamdjoug, J.R.K., Wanko, C.E.T.: Influence of artificial intelligence (AI) on firm performance: the business value of AI-based transformation projects. Bus. Process Manag. J. (2020)
48. Yin, R.K.: Validity and generalization in future case study evaluations. Evaluation **19**(3), 321–332 (2013)
49. Van Rensburg, A.: A framework for business process management. Comput. Ind. Eng. **35**(1–2), 217–220 (1998)

Chapter 5
Internet of Things Applications for Smart Environments

Kwaku Anhwere Barfi

Abstract Environmental events have a significant impact on our day-to-day existence. Changes influencing life's supplies, such as our environment, which has decreased in purity. The "smart environment" refers to a technology that provides a variety of facilities and solutions for a variety of environmental application challenges involving clean air and water. It's important to highlight that smart environment sensors combined with Internet of Things technology can offer a novel approach to tracking, sensing, and monitoring environmental objects. This has the potential to bring benefits that contribute to a greener environment and a more sustainable way of living. Environmental sensors can interact with other systems such as via Bluetooth or Wi-Fi to broadcast massive amounts of data to the network, allowing us to gain a deeper awareness of our surroundings and identify appropriate solutions to today's environmental issues. This work discusses the case for using Internet of Things to enhance e-learning in the field of environmental studies and presents a conceptualization of environmental areas of study based on Internet of Things technology. Furthermore, it analyzes a wide range of potential environmental research applications based on IoT via e-learning and e-learning platforms.

Keywords Applications · e-learning · Internet · Smart environment · Technology

5.1 Introduction

The Internet of Things (IoT) is a commonly used concept that allows users to communicate and collaborate. As it grows, it expands in scope, touching many facets of our life [1]. Kevin Ashton invented the term "IoT" while working in the Auto-ID lab at MIT in 1999, arguing that technologies like Radio Frequency Identity Devices

---

K. A. Barfi (✉)
University of Cape Coast, Cape Coast, Ghana
e-mail: kwaku.barfi@ucc.edu.gh

© The Author(s), under exclusive license to Springer Nature Switzerland AG 2022

G. Marques et al. (eds.), *Machine Learning for Smart Environments/Cities*,
Intelligent Systems Reference Library 121,
https://doi.org/10.1007/978-3-030-97516-6_5

(RFID) and Wireless Sensor Networks (WSNs) drive IoT [2]. The IoT is built on data acquired by sensors, tags, or actuators and delivered to a cloud system via a gateway. Machine-to-machine, object-to-machine, and object-to-object interactions are all feasible in the IoT [3].

From a little home to sophisticated medical equipment, the IoT offers a wide range of applications. Smart cities, smart enterprises, and smart energy utilization are just a few of the human activities covered by the IoT. One of the most visible human activities influenced by the IoT is education, which has the potential to turn education into a more imaginative framework in the near future [4]. The IoT is a relatively recent phenomenon that fosters innovation in a variety of industries. One of these is the field of education (e-education). Because IoT may be combined with other information technologies (IT), it can provide a wide range of e-educational technologies that have the potential to change education systems in the future. Smart things will be integrated into the future teaching center. Students and teachers must provide their fingerprints and RFID ID cards to the reader, as well as mobile verification, to gain access to the school's physical rooms or access to the school's automated system management. In the future, sensors will be used in IoT classrooms to authenticate educators' and students' access.

The IoT has the needed strength to link all physical and virtual objects in a secure manner. This will allow students to connect to labs and libraries -online- for productive online learning activities. Online learning is now usable through the rapid growth of information and communication technologies (ICTs), as well as the availability of the Internet. By combining real and virtual aspects and interfering with the learning process, the IoT will allow for the expansion of learning contexts. The introduction of the e-learning concept for use in the virtual environment can give learners virtual access, but it has its drawbacks. The primary constraints in e-learning include geographical location, technical issues, and communication between virtual users and agents. One solution to the aforementioned challenges is the use of smart objects in the learning environment. In e-learning ecosystems, the IoT is widely regarded as the most important source of intelligent agents [5].

Smart campus are examples of IoT characteristics that can be added to traditional e-learning [6]. By linking a large number of distant learning devices and things. A high level of interaction between virtual and physical elements can be employed to create a variety of collaborative settings [5]. As previously said, the IoT application's scope has expanded to incorporate complex healthcare systems as well as basic smart homes, smart connected cities, and urban system management.

All large human daily activities are included in the IoT, including e-learning, which has recently become a focus of urban smart city academics. In the development of smart cities, the IoT has become a vital component. It is used to transition from a knowledge transfer to a collaborative e-learning paradigm in order to improve e-citizen knowledge and develop their capacities in the e-community [7]. By leveraging the most advanced technology, urban IoT is created to implement smart city ideas. The integration of diverse technical technologies would make cities smarter, according to one of the many components of the smart city concept. Integrating the IoT into ICTs might help improve administrative procedures by providing more value-added services and educational opportunities. Smart cities can use a number of IoT initiatives to reduce environmental issues and improve educational platforms and services. The objective of this chapter is to throw more light on the applications of environmental research based upon IoT through electronic platforms. This chapter will also talk about IoT in education, IoT benefits on e-learning, smart collaboration and smart homes.

5.1.1 IoT in E-Education

Education is a foundational concept in today's human activities [8]. The application, network, and perception layers make up the fundamental IoT architecture [9]. Through an interface, the application layer delivers services to end-user applications. The network layer guarantees that the nodes and the gateway are connected. Through the gateway point, data from sensor nodes in the perception layer is captured and sent to a cloud system. It's crucial to understand that the sensing layer includes tangible items or sensors capable of sensing an event or object's behavior. A sensor system with limited capability is included in this layer. The sensors in this system collect data and manage RFID and WSN. Figure 5.1 shows an IoT educational framework in which IoT devices recognize events, after which they can track objects in the perception layer. The information gathered by the senses will be sent to a gateway and stored in a tiny cloud system. When the data is processed, it can be used to make more judgments.

5.2 IoT Technology Benefits on E-Learning

The educational system has changed as a result of modern Internet technology. Ubiquitous technology has improved the education industry in a variety of ways, including

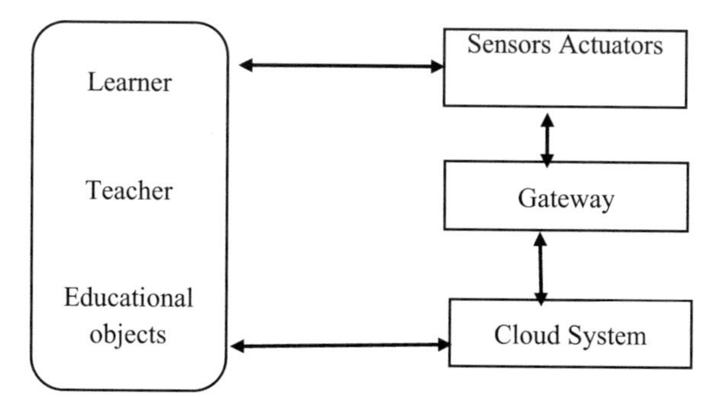

Fig. 5.1 IoT educational framework

the ability to teach and learn. It refers to the Internet's and related technology's high level of accessibility, which has converted earlier educational approaches into modern educational concepts, especially in the field of e-learning. Through IoT activities, students can interact with teachers. Physical labor is no longer a priority for pupils or teachers. Instead, they may focus on the most important part of a student's learning efficiency or learning activities. They can collect data on learner efficiency using technology [1]. On a daily basis, the internet has an impact on people's lives [1]. It has a lot of benefits in a variety of learning-related fields. The adoption IoT can allow teachers and students improve upon their performance in the online learning environment.

5.2.1 Online Self Learning

Self-learning, or autodidacticism, is founded on the idea of encouraging students to read literature without the help of an instructor. Students who study this way are self-motivated to complete their homework on their own time [5]. Because objects (people) could not interact with other objects before the IoT, they may now assist humans in acquiring information on their own. Consider a scenario where a few touchable wireless personal computers (PCs) are strategically placed throughout a university campus. When a student has a question or a problem, he or she can turn to the computer for assistance. By completing an authentication process that includes physical scanning, the user can gain access to the system and obtain all necessary information (fingerprint or RFID card). Students can access a wide range of educational resources from anywhere, at any time, while performing homework, gathering materials, and submitting and receiving homework from instructors. The IoT allows you to link any online educational resources for information at any time and from any location. By registering their cellphone for physical identification, students can connect their devices (as previously indicated) to conduct experiments and retrieve whatever information they require. They can complete exams and assignments on time, send work, and receive exam results.

5.2.2 Smart Collaboration

The availability of an IoT-enabled online environment enhances smart collaboration or communication among users. Communication is critical for educational infrastructures that have collaborative objectives. A high level of communication and interactions is generated due to the integration of physical and virtual items as well as the extensive use of the IoT [1]. From a technological standpoint, the IoT can offer a large-scale and efficient collaboration platform using smart technologies like RFID and WSN. When choosing a platform for collaboration, be aware that not all are intelligent. Collaboration can be transformed into various high-performance smart collaborations by utilizing contemporary Internet technologies that support the IoT. Smart collaboration is created by the concept of IoT. In a smart collaborative setting, a better schedule for activities can be organized for the assistive team's collaborative solution [8].

It is thought that creating a smart city leads to effective and efficient learning. The use of IoT in smart cities allows engagement among learners. IoT deployment in smart cities promises to have a substantial impact on learners and instructors. There could be important instruments for supporting IoT-based learning systems, which, despite existing issues, are creating more efficient techniques in the smart city's e-learning ecosystem. In light of this reality, future study will focus on existing e-learning systems and their compatibility with smart cities.

5.3 Internet of Things and Smart Homes and E-learning

The IoT ushers in a new era in which all of our electrical equipment is connected by a network [8]. The IoT is also redefining the concept of a "smart house." Smart home with IoT enabled convenience for household members. By 2022, the worldwide smart home market is estimated to have grown to USD 119.26 billion [8]. In the view of [8], companies are making items to make users feel at home concepts by offering cutting-edge smart home services.

Smart home adoption is hampered by a number of issues, including high device prices, low client demand, and extended equipment replacement cycles [10]. The most significant stumbling block is a dearth of technology to build the infrastructure of a smart home [8]. In the view of [1], smart house opinions have failed to appreciate potential consumers' genuine needs from a smart home from technology or technical standpoint.

5.3.1 Smart Homes

A smart home is more than just a place with all the latest technology, but a network of communication technologies, high-tech household equipment, appliances, and sensors that are accessible, monitored, and controlled remotely, with services that adjust to the needs of the people [11]. Although it was possible for the home network industry to thrive in the late 1990s because of the use of high-speed Internet, smart homes did not emerge until much later, when the use of smartphones became widespread. People have used the terms smart house, intelligent home, home automation, and home network to describe the same concept. With the ideas of IoT and the presence of a smart home with internet connectivity, there has been a noticeable rise in the use of smart homes since the middle of the 2010s.

A smart home is a more advanced type of home automation. It is a common use of communication technology that is implemented in order to integrate with various services to deliver secure, economical, and comfortable operation of the home [11]. This is why smart home services like smart lighting and heating controls were deployed to control environmental systems [1]. A smart home service currently monitors user actions and the internal environment at home, thanks to technological advancements. In addition to offering services that are customized to the user's desires and needs, a smart home also offers personalized services.

As artificial intelligence (AI) has become increasingly ubiquitous, smart home services have evolved. Amazon Lab126's intelligent personal assistant "Alexa" has been integrated into a wide range of goods. Alexa has been integrated into LG Electronics' smart home product line. By saying "Alexa" from a smart refrigerator, for example, a user can access services such as news searches, shopping virtually, and appointment checks. Xiaomi has released a smartphone that can be plugged into any equipment, including refrigerators, air conditioners, and washers. Apple is working on an artificial intelligence speaker that will enable "Apple HomeKit," which would let users to operate their home kit gadgets with voice commands. As a result of the application of IoT and AI, smart home services are growing and expanding.

Prior smart house research has taken a technical or limited approach. Bayani and Vílchez [1], for example, created a smart floor technology for a smart home that tracks who, where, and what users are doing. Bayani et al. [12], demonstrated a bracelet that tracks users' daily activities at home. Andoh et al. [13] presented biometrics monitoring system for measuring pulse and respiration rate that may be utilized at home. Only young nontechnical professionals were used by Koskela and Väänänen-Vainio-Mattila [14]. Sun and Shen [8] performed research to see if customers appreciated smartness through common communication gadgets including computers, televisions, and cellphones. To test inhabitants, [1] created an autonomous energy management system. These investigations have greatly aided in improving the comprehensiveness of a smart home solution. However, because there is a scarcity of research on business and consumer viewpoints, additional study is needed to stimulate this development.

5.3.2 Smart Home Service Adoption

The term "smart" has been used in a variety of contexts, including smartphones, smart TVs, smart learning, and smart homes. Despite the fact that each name has its unique meaning, they all mean "intelligent," which is a low-level version of AI. However, it remains to be seen whether such an approach can be leveraged to construct an intelligence service capable of totally replacing human decision-making. Because of their desire for freedom, uncertainty, and suspicion of technology, humans will be unwilling to grant machines entire decision-making authority. People's desire for intelligence varies according to their individual features and circumstances. Some people are concerned about objects that are intelligent and sophisticated. When AlphaGo defeated a human in the Go game, for example, some people expressed reservations about AI, claiming that a machine may control or kill people. As a result, contrary to common opinion, "smart" may refer to a limited range of intelligence that may be controlled by humans.

Since the mid-2000s, researchers have been looking into the acceptability of smart home services. The technology acceptance model (TAM), also known as the unified theory of technology acceptance and use, has been applied to specific populations such as the elderly, disabled, and patients in the majority of studies (UTAUT). Self-efficacy has a critical part in the acceptability of a smart home for the elderly, according to Leeraphong et al. [15]. Patients' quality of life expectations was found to have a substantial impact on smart home adoption in the [16] study. Existing hypotheses have been expanded while new factors have been explored. [17] stressed the importance of continuing to develop AAL (ambient aided living), which assists disabled and elderly individuals, especially those with chronic illnesses, in improving their well-being and enabling independent living. The adoption of telecare systems, a sort of smart home service, was investigated, and it was discovered that perceived utility plays a role in the device's desire to be utilized [18].

As smart services become more widely used, there has been a spike in user acceptance studies. Bao et al. [19] investigated the mobile smart home environment and found that compatibility and safety are important considerations when selecting whether to adopt. According to Park et al. [20], the futuristic notion of smart home adoption with IoT technologies would extend by presenting wireless network accessibility, as well as compatibility with a variety of operating systems, languages, and frameworks. Furthermore, the study discovered that aspects including fun, compatibility, connection, control, and cost influence smart home acceptability. Controllability is crucial for adopting a smart home, according to Ji et al. [21]. Furthermore, intelligent control may automatically learn user behavior and demand, enabling more intelligent appliance control.

Security and privacy, according to several studies, are critical aspects of smart home acceptance. For example, [22] discovered that deploying a smart home service improves family security. Specific privacy and data protection, according to Elmasllari and Al-Akkad [23], are required to operate smart concepts. According to Banerjee et al. [24], safety issues that affect user behavior can be minimized by adopting flexible security policy enforcement mechanisms.

Because the current disagreements are abstract, a fundamental grasp of smart home services is required to categorise them. Furthermore, previous research focused on a single user group, but with the widespread adoption of smart home services, more in-depth talks for a variety of user groups are now required. This can help students who adopt distance learning, e-learning and mobile learning to get more teaching materials related to their studies. The application smart home service adoption can support e-learning and e-learning platforms.

5.3.3 *Critical Factors for Smart Home Service*

The first smart home service was pushed through domestic system automation with the goals of ease, comfort, stability, amenity, health, household labor reduction, and energy efficiency. The definition of a smart home has now been expanded to encompass services that can be operated remotely at any time and from any location thanks to advancements in wireless Internet and smartphones. Household electrical appliances, information and communication devices are all connected on the IoT age, and the smart home is evolving into a type of AI service that self-learns the behavior of its residents. As a result, in the IoT era, the smart home is a concept that combines connectivity with traditional elements like automation and remote controllability. Issues such as service dependability, security, and privacy, according to some, may hinder users' acceptance. A combination of these factors can be used to describe the service's dependability. The smart home environment is a factor that must be explored since it is intricately linked to the user's life and could cause substantial harm in the event of a risky circumstance. As a result, while deciding whether or not to employ a smart home service, automation, remote controllability, interconnection, and dependability are all important elements to consider.

Automation is defined as a machine agent (typically a computer) performing a previously conducted by a human function [25]. The automation of households and home infrastructure was one of the early smart home's main goals, and home automation was the initial name for the smart home service. In recent years, automation has gained pace as information technology has made it more cost-effective and simpler to implement. Higher-level automation is now possible, thanks to a spike in AI interest in recent years. AI technology can improve the operation of a smart home by intelligently helping people. As a result, technology is one of the most important aspects of a smart house.

A smart house has the advantage of being able to be managed remotely via mobile devices. Users want to be able to access smart home services as soon as possible, therefore this is a crucial component of a smart home system. However, you'll need a network connection to create an intelligent and remote-controllable smart home system. Bluetooth, Z-Wave, and Wi-Fi are just a handful of the networks that are available. Networks should be standardized and integrated to enable remote operation in order to make smart home services more accessible. The Wi-Fi protocol is used by the majority of electrical items, allowing mobile devices to operate household appliances. When the remote control is available, the complete concept of smart can be deployed anywhere and at any time.

The ability of devices, apps, and services to interact and function together is referred to as interconnectedness. In order for a smart home to expand, equipment must be able to adapt to changes in a user's preferences, requirements, and wishes. A smart home's system should be simple to connect to new devices. In order to function properly, a smart home must communicate with other smart homes on the network. Many forms of network and communication protocols, on the other hand, constitute a real-world impediment. There are wired and wireless networks, as well as a variety of communication technologies, to choose from. Due to the high cost of satellite communications and the limited transmission between electronic devices, the technical standard is presently insufficient [25].

Controllability, interconnectivity, and reliability have been shown to have a substantial impact on smart home service acceptance behavior in studies. It's amazing that automation has had no noticeable effect. Remote management features are chosen over highly advanced automated services because they are safer and more effective, as evidenced by the following facts: Because their home is safe and represents their personal zone, people may choose to have control over their smart home's equipment rather than having it totally automated.

Furthermore, group comparison analysis yields a diverse set of results. Controllability and connectedness are important to apartment inhabitants, whereas automation and reliability are important to traditional home dwellers. Differences in infrastructure between apartments and single-family homes could be a factor. Smart home services have recently been installed in new homes in Korea (remote heating management, gas shutdown, etc.). As a result, it appears that apartment occupants desire more comprehensive and precise supervision. Because most homes lack networked and automated capabilities for household control, residents may seek the automation and dependability of smart home services.

Male's place higher importance on connectivity than females do, whereas females value reliability more. The age-related comparisons yielded similar results. This is due to the fact that women and the elderly are more likely to avoid taking risks. It's logical that they prefer these qualities to men or young individuals since they desire stability. Men and young people demand fascinating and unique services, which the interconnection of smart homes may help meet. Users of similar services have emphasized the need for automation and consistency. Even individuals who disdain control and interconnection are likely to have encountered instances in which the control and connectedness of prior systems did not guarantee usability.

The rapid development and implementation of smart and IoT based technologies have allowed for various possibilities in technological advancements for different aspects of life. The application of IoT is to ensure a better efficiency of systems (technologies or specific processes) and finally to improve life quality. IoT technologies are an opportunity for humanity to provide a wide range of e-educational technologies that have the potential to change the educational systems. This allows students to connect to labs and libraries online for productive learning activities using smart devices. Online learning is now usable through the rapid growth of information and communication technologies, as far as there is availability of the Internet. By combining real and virtual aspects and interfering with the learning process, the IoT will allow for the expansion of learning contexts. The availability of an IoT-enabled online environment enhances smart collaboration or communication among students. Students can collaborate with their peers and share assignment-related information. This chapter considered the IoT in education, IoT technology benefits on e-learning, online self-learning, smart collaboration, Internet of things and smart homes and e-learning, smart homes, smart home service adoption and critical factors for smart home service.

5.4 Conclusion

The content of this report is relevant to those individuals who plan to create or market smart home services with the support of IoT. Functional diversity must be provided first and foremost to consumers. It is important to guide customers through the process of selecting the functions they desire, as every customer group has different smart home service needs. For example, while providing a smart home service for apartment residents, additional devices can be used to provide additional automation functions for users of ordinary houses. In order to assist with that, smart collaboration, smart home and smart home service adoption providers should partner with third-party device manufacturers and develop numerous different plan configurations for various customers, including serving clients by collaborating with other device makers. Additionally, research and development on AI-enabled automation must be constantly ongoing. Automation was discovered to have a positive impact on current smart home consumers' intent to keep using their products permanently. As a result, it's reasonable to predict that demand for automation will increase as basic controllability-based smart home services become more widely available. Additionally, in order to increase the level of trust clients have in the services they receive, smart home service providers should use the most secure security technology available with the concept of IoT and establish strict internal procedures to prevent data leakage. With data-driven smart home companies like Google rapidly growing, smart home service providers' trust has become a big worry.

This chapter gave an introduction on IoT in education, IoT benefits on e-learning, smart collaboration and smart homes. Further study can attempt to gather diverse data to ensure a better understanding of smart home services adoption in both developed and developing countries. Getting the right understanding of IoT can help improve the quality of life of students and enterprises production opportunities. Moreover, this can bring a challenge to IoT in higher education.

References

1. Bayani, M., Vílchez, E.: Predictable influence of IoT (Internet of Things) in the higher education. Int. J. Inf. Educ. Technol. **7**, 914–919 (2017) https://doi:https://doi.org/10.18178/ijiet.2017.7.12.995
2. Wood, A.: The Internet of Things is Revolutionizing Our Lives, But Standards are a Must. The Guardian, UK (2015)
3. GSMA Association.: Understanding the Internet of Things. https://www.gsma.com/2014
4. Maksimović, M.: Transforming educational environment through green Internet of Things. Trendovi Razvoja, Paper No. T1.1-3. 1-4 (2017)
5. Soava, G., Sitnikova, C., Danciulescu, D.A.: Optimizing quality of a system based on intelligent agents for e-learning. In: 21st International Economic Conference, May, Sibiu, Romania, 47–55 (2014)
6. Jafari, A.: Conceptualizing intelligent agents for teaching and learning. Educ. Q. **25**, 28–34 (2002)

7. Zanella, A., Bui, N., Castellani, A., Vangelista, L., Zorzi, M.: Internet of things for smart cities. IEEE Internet Things J. **1**, 1–10 (2014)
8. Sun, G., Shen, J.: Towards organizing smart collaboration and enhancing teamwork performance: a GA-supported system oriented to mobile learning through cloud-based online course. Int. J. Mach. Learn. Cybern. **1**, 391–409 (2016)
9. Sethi, P., Sarangi, S.R.: Internet of things: architectures, protocols, and applications. J. Electr. Comput. Eng. **9**, 25 (2015). https://doi.org/10. 1155/2017/9324035
10. Edwards, W.K., Grinter, R.E.: At home with ubiquitous computing: seven challenges. In: Abowd, G.D., Brumitt, B., Shafer, S., (eds.) Ubicomp 2001: Ubiquitous Computing. Berlin, Heidelberg, 2001, Proceedings, pp 256–272. Springer International Publishing (2001). https://doi.org/10.1007/3-540-45427-6_22
11. Chen, Y., Yang, M.: Study and construct online self-learning evaluation system model based on AHP method. In: 2nd IEEE International Conference on Information and Financial Engineering, 54–58 (2010)
12. Bayani, M., Segura, A., Saenz, J., Mora, B.: Internet of things simulation tools: proposing educational components, SIMUL, Greece, Athens. In: The 9th International Conference on Advances in System Simulation, IARIA, 57–63 (2017)
13. Andoh, H., Watanabe, K., Nakamura, T., Takasu, I.: Network health monitoring system in the sleep. In: SICE 2004 Annual Conference, Sapporo, Japan, pp. 1421–1424. IEEE (2004)
14. Koskela, T., Väänänen-Vainio-Mattila, K.: Evolution towards smart home environments: empirical evaluation of three user interfaces. Pers. Ubiquit. Comput. **8**, 234–240 (2004)
15. Leeraphong, A., Papasratorn, B., Chongsuphajaisiddhi, V.: A study on factors influencing elderly intention to use smart home in Thailand: a pilot study. In: The 10th International Conference on e-Business, Bangkok, Thailand, 1–10 (2015)
16. Alaiad, A., Zhou, L.: Patients' behavioral intentions toward using WSN based smart home healthcare systems: an empirical investigation. In: 48th Hawaii International Conference on System Sciences, Kauai, 824–833 (2015)
17. Fan, X., Xie, Q., Li, X.: Activity recognition as a service for smart home: ambient assisted living application via sensing home. In: IEEE International Conference on AI & Mobile Services (AIMS), Honolulu, 54–61 (2017)
18. Sebastiaan, T.M., Eveline, J.M., Joost, H., Katrien, G.L., Hennie, R.B., Hubertus, J.M.: Factors influencing acceptance of technology for aging in place: a systematic review. Int. J. Med. Inform. **83**, 235–248 (2014). https://doi.org/10.1016/j.ijmedinf.2014.01.004
19. Bao, H., Chong, A.Y.L., Ooi, K.B., Lin, B.: Are Chinese consumers ready to adopt mobile smart home? An empirical analysis. Int. J. Mobile Commun. **12**, 496–511 (2012)
20. Park, E., Cho, Y., Han, J., Kwon, S.J.: Comprehensive approaches to user acceptance of Internet of Things in a smart home environment. IEEE Internet Things J. **4**, 2342–2350 (2017)
21. Ji, J., Liu, T., Shen, C.: A human-centered smart home system with wearable-sensor behavior analysis. In: IEEE International Conference on Automation Science and Engineering, Fort Worth, 1112–1117 (2016)
22. Tanwar, S., Patel, P., Patel, K., Tyagi, S., Kumar, N., Obaidat, M.: An advanced Internet of Thing based security alert system for smart home. In: International Conference on Computer, Information and Telecommunication Systems, Dalian, China, 25–29 (2017)
23. Elmasllari, E., Al-Akkad, A.: Smart energy systems in private households: behaviors, needs, expectations, and concerns. In: IEEE 14th International Conference on Networking, Sensing and Control, Calabria, Italy, 52–157 (2017)
24. Banerjee, M., Lee, J., Choo, R.: A blockchain future for internet of things security: a position paper. Digit. Commun. Netw. **4**, 149–160 (2018). https://doi.org/10.1016/j.dcan.2017.10.006
25. Bayani, M., Marin, G., Barrantes, G.: Performance analysis of sensor placement strategies on a wireless sensor network. In: IEEE 4th International Conference on Sensor Technologies and Applications, Sensorcomm, 609–617 (2010)

Chapter 6
Exploring Interpretable Machine Learning Methods and Biomarkers to Classifying Occupational Stress of the Health Workers

Analúcia Schiaffino Morales, Fabrício de Oliveira Ourique, Laura Derengoski Morás, and Silvio César Cazella

Abstract Occupational stress is present in all professions but poses a high risk to health professionals' performance when it becomes bad stress (distress). This chapter presents issues related to monitoring health professionals' stress through wearable devices, focusing on research into the use of biomarkers and machine learning techniques used in the development of models that can help in decision-making related to coping with distress. Results from a literature review on these topics are reported. Challenges related to explainable Artificial Intelligence are addressed in the chapter, and challenges associated with the definition of a stress classification at different levels seek that, to identify the impacts on the health of the health professional. As the main contribution, this chapter presents a proposal for an intelligent system, the result of ongoing research, which helps in recommending actions according to the level of perceived stress, in an explainable, transparent way that is reliable for adoption by these professionals' managers. Challenges related to the intelligent system implementation itself, such as the security and privacy of user information in the foreseen layers of the Health Internet of Things solution architecture, are described for reflection.

A. S. Morales (✉) · F. de Oliveira Ourique
Computer Department (DEC), Sciences, Technologies and Health Education Center (CTS), Federal University of Santa Catarina (UFSC), Florianópolis, Brazil
e-mail: analucia.morales@ufsc.br

F. de Oliveira Ourique
e-mail: fabricio.ourique@ufsc.br

L. D. Morás · S. C. Cazella
Department of Exact Sciences and Applied Social (DECESA)—Graduation in Information Technology and Healthcare Management, Federal University of Health Sciences of Porto Alegre (UFCSPA, Porto Alegre, Brazil
e-mail: lauradm@ufcspa.edu.br

S. C. Cazella
e-mail: silvioc@ufcspca.edu.br

© The Author(s), under exclusive license to Springer Nature Switzerland AG 2022

G. Marques et al. (eds.), *Machine Learning for Smart Environments/Cities*, Intelligent Systems Reference Library 121,
https://doi.org/10.1007/978-3-030-97516-6_6

Keywords Biomarkers · Occupational stress monitoring · Explainable artificial intelligence · Healthcare internet of things

6.1 Introduction

For about a year, health professionals who work on the front line have been facing excessive working hours. This scenario involves professionals worldwide, and even with the advance of immunization, new variants have emerged affecting their working hours. Problems related to occupational activity in the face of a pandemic are reported by the World Health Organization (WHO), among which risks to health professionals stand out, such as exposure to pathogens, long working hours, psychological distress, fatigue, professional burnout, stigma and physical and mental violence. Stress in excessive doses can have undesirable consequences, such as fatigue, irritability, depression, lack of concentration. It can negatively interfere in the work environment and, consequently, decrease the productivity of individuals [1]. The stressful conditions to which health teams have been subjected around the world can lead to security failures in the correct use of Personal Protective Equipment (PPE) in the care of infected patients, putting the health of the professional at risk, as well as of your team, and your family members [2, 3].

The editorial in Occupational Medicine points out that health professionals are considered the groups at most significant risk in facing COVID-19 [4]. Physical signs of stress can be seen in the form of rapid heart rate, heavy breathing, excessive sweating, stomach pain, dry mouth, tremors, and cold extremities, or manifest in mind through signs such as insomnia, insecurity, anguish, fear, hopelessness, panic attacks, among others. Decreased productivity, work accidents, and even increased health costs and absenteeism at work are some of the consequences due to negative stress in the occupational environment [5]. Known as distress, negative stress can affect work activities and the quality of life of these professionals who are more exposed due to the pandemic [6]. Given the importance of the subject, there are several studies published in the scientific literature regarding identifying stress through the use of biomarkers [7].

With the advancement of information and communication technologies, as well as electronic devices, the Internet of Things (IoT) has been leveraging a series of researches in the areas of wearable sensors and biomarkers. Ways of capturing individuals' reactions through the use of non-invasive sensors have been the subject of investigation. However, a single device that combines individual information to monitor occupational stress has not been found.

In the scientific literature [1–3], a scale for occupational stress categories for health professionals has not been found. A study on biomarkers that can help in the investigation and identification of occupational stress is presented. In this way, this chapter suggests essential biomarkers that can be observed to classify different levels of stress of health professionals.

In the development of the chapter, important explainable characteristics for the context of occupational stress are presented, and a study on the biomarkers that can help in its investigation and identification. Furthermore, how explainable algorithms can help to determine this scale, noting that in recent years, society as a whole has encouraged the use of machine learning algorithms that are not black-box solutions [8]. This applies in areas where there is a need to justify the reliability of systems based on these algorithms. The lack of transparency in decision-making in areas involving decisions about human lives reduces the reliability of systems, and wrong choices can have catastrophic consequences. This applies to the medical field, diagnostic processes, anomaly identification, treatment recommendations, and other applications need to be highly reliable to assist in decision-making processes and to be accepted by the medical profession. Furthermore, there are already techniques that help validate a machine learning algorithm [8–10].

This chapter is organized into seven sections, including the introduction and the final considerations. A literature review on biomarkers for the area of occupational stress is presented, considering wearable IoT resources and machine learning algorithms. A proposal of different biomarkers and occupational context to identify different levels of stress in health professionals. Then, characteristics related to the explainability of a recommendation system applied to occupational health are presented. And it ends by pointing out the importance of increasing reliability through the use of information security and privacy features in the most diverse layers for the Internet of Things that are applied to health.

6.2 Biomarkers and Machine Learning Models

Aiming to report the biological markers used for stress detection, a literature review was performed. The implemented review protocol focused on analyzing biomarkers collected from wearable sensors as input information for Machine Learning models for stress recognition.

This review intends to answer the three following research questions:

1. What are the biological markers collected from the research participants for stress recognition?
2. Which non-invasive device/sensor is used to collect these biomarkers?
3. What machine learning algorithms were implemented to analyze and/or predict the collected features to build a stress recognition system?

Before the search process, the strategies and parameters were determined to get the most significant number of articles related to the subject. The following items describe these decisions:

- *Keywords*: biomarker AND stress AND wearable AND 'machine learning';
- *Publication period*: four years, from 2017 up to 2021;

Table 6.1 Inclusion and exclusion criteria

Inclusion criteria	Exclusion criteria
Original articles	Review articles
Peer-reviewed articles	Duplicate articles
Articles written in English, Spanish or Portuguese	Articles with invasive data collection
	Articles without the use of machine learning or deep learning models
	Articles describing the recognition of other conditions, and not stress

- *Search Engines*: IEEE Xplore,[1] Scopus,[2] Science Direct,[3] Pubmed Central,[4] and Bireme.[5]

The search for articles published in the last four years using consolidated search engines in medicine and computer science tends to guide the results to the state-of-the-art machine learning models used for stress recognition. However, the operational differences between these search engines do not allow the use of complex filters that can be applied equally for all these databases. Thus, creating inclusion and exclusion criteria is needed to refine the results in a preliminary analysis. These criteria are presented in Table 6.1.

Articles metadata information such as title, author, and abstract was downloaded in electronic formats (CSV, XML) and inserted in Google Spreadsheet. After this step, articles were filtered based on the inclusion and exclusion criteria. At the end of this stage, the 13 remaining articles were downloaded in PDF format and read. During the reading stage, items such as primary objectives, biomarkers, wearable sensors, machine learning algorithms, and evaluation metrics were collected and registered in Google Spreadsheet for later analysis.

A summary of the number of articles found and analyzed is shown in Fig. 6.1, as the PRISMA Flow Diagram [11].

Besides the articles retrieved from the search engines, three more were included in the analysis process, found through other sources. The number of articles retrieved, per database, is shown in Table 6.2.

Table 6.3 summarizes the findings related to the review subject, aspects such as:

- *Biomarker*: the biological marker collected from the participants, related to stress recognition;
- *Wearable*: the non-invasive device/sensor used to collect the biomarkers for stress;
- *Machine Learning (ML)/Deep Learning (DL) model*: the algorithm(s) used to analyze the collected features and build a stress recognition system.

[1] https://ieeexplore.ieee.org/Xplore/home.jsp.

[2] https://www.scopus.com/.

[3] https://www.sciencedirect.com/.

[4] https://pubmed.ncbi.nlm.nih.gov/.

[5] https://bvsalud.org/.

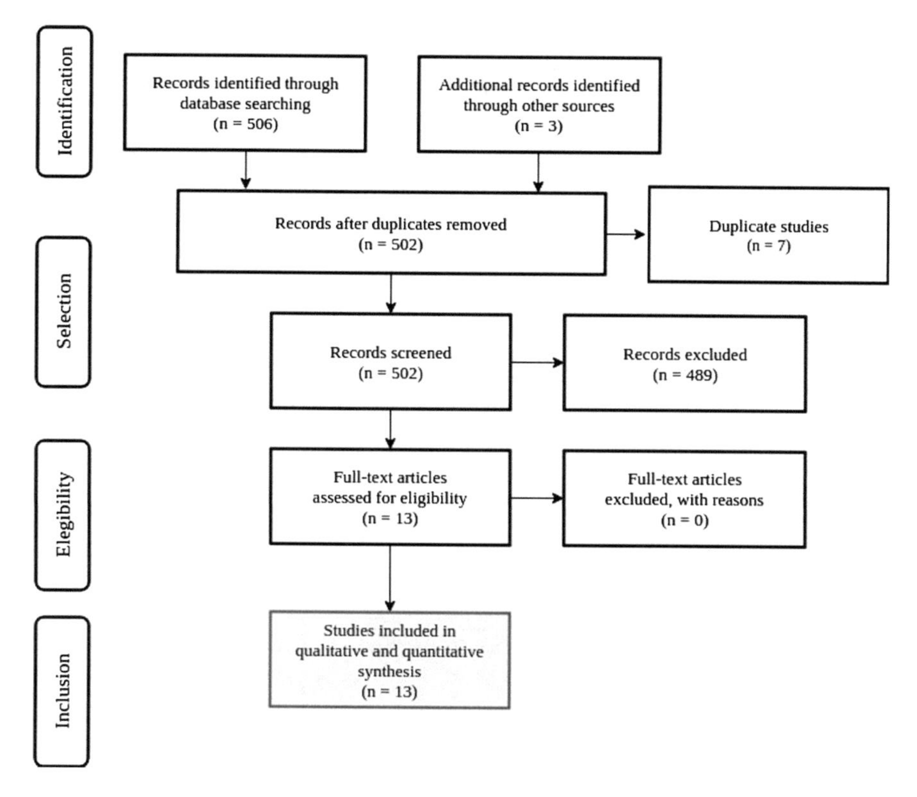

Fig. 6.1 PRISMA flow diagram for the literature review

Table 6.2 Number of articles retrieved per search engine

Database	Retrieved	Duplicate	Removed	Included
Science direct	486	0	483	3
Scopus	9	0	3	6
PubMed	5	4	1	0
Bireme	2	0	2	0
IEEE Xplore	4	3	0	1
Total	506	7	489	10[a]

[a]The total of articles refers only to those retrieved from the search engines, three more were included from other sources

The following abbreviations are used to refer to some terms in the articles included in the review: electrodermal activity (EDA), heart rate (HR), skin temperature (ST), blood volume pulse (BVP), Electrocardiogram (ECG), Electromyogram (EMG), Body Temperature (TEMP), Respiration (RESP), Three-Axis Acceleration (ACC), Galvanic Skin Response (GSR), heart rate variability (HRV), Electroencephalogram

Table 6.3 Summary of the data collected from the reviewed articles

Ref	Biomarker	Wearable	ML/DL model
[12]	EDA, HR, accelerometry, and ST	Wristwatch commercial device	25 classification models (DT, DA, LR, NB, SVM, NN, and EC)
[13]	BVP, ECG, TEMP, EDA, EMG, RESP and ACC	Wristwatch and chest-worn commercial devices	CNN
[14]	ECG, GSR, TEMP, and BVP	Wrist-worn prototype device	Feed-forward neural network
[15]	Pulse wave, speech, EDA	Wrist-band commercial device	SVM, DT, and DL algorithms
[16]	ECG, BVP, ST, RESP, and EDA	Wristwatch and chest-worn commercial devices	kNN, DT, and RF
[17]	Time and frequency metrics derived from HRV	Wristwatch commercial device	kNN, SVM, MLP, RF and GB
[18]	EDA and BVP	Wrist-band commercial device	RF, k-NN, LR, and SVM
[19]	EDA	Wristwatch commercial device	SVM and Deep-SVM
[20]	ST, EDA, HRV, and ACC	Wristwatch commercial device	MATLAB classification learner application: DT, DA, LR, NB classifiers, SVM, NN classifiers, and EC
[21]	HRV, EDA, and EEG	Wrist-band, headset, and electrodes commercial devices	SVM
[22]	ECG, BVP, EDA, EMG, RESP, TEMP and ACC	Wristwatch and chest-worn commercial devices	FCN, Resnet, MLP, time-CNN, MDCNN, STRESNET, CNN-LSTM, MLP, MLP-LSTM, inceptiontime
[23]	EEG	Electrodes system prototype	LR, SVM, and NB
[24]	ECG and RSP	Sensor-based wearable device	RF and SVM

(EEG); Deep Learning (DL), Decision Trees (DT), discriminant analysis (DA), logistic regression (LR), naïve Bayes (NB) classifiers, support vector machines (SVM), nearest neighbor (NN) classifiers, ensemble classifiers (EC), Convolutional Neural Network (CNN), Random Forest (RF), Gradient Boosting (GB), Multilayer Perceptron (MLP), Fully convolutional network (FCN), Residual network (Resnet), Time convolutional neural network (Time-CNN), Multichannel deep convolutional neural network (MCDCNN), Spectrotemporal residual network (Stresnet), Convolutional neural network with long-short term memory (CNN-LSTM), Multilayer perceptron with long-short term memory (MLP-LSTM).

It is noteworthy that the non-invasive forms have been used, as seen from the data in Table 6.3, as promising for identifying stress. Among the non-invasive parameters that have shown greater reliability, the following stand out: heart rate rates, both HR and electrocardiogram, measurements of skin sweat and skin changes, values related to breathing and body temperature. Other biomarkers show changes, physical and physiological, due to the functioning of the sympathetic nervous system that increases the level of hormones, changing homeostasis and causing changes in the body as a whole, such as a pupil diameter, blood pressure, change in the musculature such as muscle contractions, glucose level, increased cortisol and changes in breathing.

6.3 Healthcare Internet of Things

Directly related to machine-to-machine systems (M2M, Machine to Machine), which encompass telecommunications networks, which include access devices to transmit data through remote applications, to monitor, measure, and control the device, the environment around it, or even the data systems connected to it through networks [25]. Internet of Things (IoT) is a term used to represent a set of technologies, communication protocols, systems, and resources needed to adapt objects or things to the Internet network. There are several areas interested in this new concept. The Healthcare Internet of Things (HIoT) stands out in this chapter. There is no single definition for this area in the scientific literature with different nomenclatures: Smart Healthcare, Internet of Medical Things, Hospital 4.0, Smart Hospital, etc. However, there is a consensus regarding the need for a layered organization, being more common to find: Things, Processing, and Cloud suggested in most scientific articles [26]. Several benefits are pointed out for adopting IoT in healthcare, highlighting cost reduction, improved service, efficient management, international collaboration, and personalized service. It is still considered challenging in terms of scalability, interoperability, the volume of information, and especially privacy and information security [27].

However, it is noteworthy that areas whose systems decisions involve human lives, such as health, need special care. Research to employ machine learning in medical diagnostics, vital signs monitoring, elder monitoring, and other applications requires a high level of reliability. Systems need to present a path to assist in making medical decisions. HIoT has shown promising results, boosting interest in the area of sensor devices, increasing the number and types of communication protocols, providing platforms for data acquisition, proposing information processing through intermediate layers, distribution, security, and synchronism [28]. So that these systems can be adopted in healthcare organizations, improving the patient management and monitoring system, or other applications. It is necessary to add interpretability to the processing of information [10].

6.3.1 *Heart Activity, Skin Response, and Temperature*

The variation in heartbeats has been considered an efficient non-invasive measure to detect cardiovascular conditions. Five of the related works use ECG, and four use the heart rate variability, as shown in Table 6.3 in the previous section, presented in the review. These variations are related to the activities of the autonomic nervous system (ANS) and, therefore, are considered an adequate parameter to monitor stress according to the scientific literature [29]. The variation in heart rate reflects the ability of individuals to adapt to changes, which is related to the activities performed by work activities, as in the case of health professionals facing a pandemic. Heart rate variability can be measured either by optical sensors or through ECG (Electrocardiogram). The ECG provides a graphic record of the electrical activity produced by the cardiac muscle, and it is possible to use it in wireless wearable devices as presented in [30]. In recent years, in some devices used for cardiac monitoring during physical activities, measurement by optical sensors has been used, which work through light beams that cross the skin layers and measure changes in blood volume in the pulse area, which is more suitable for a wearable to identify occupational stress due to its ease of use during the most activities. Additionally, the mean heart rate, some features from ECG pointed at [31] and ought to be employed to classify the distress. For example, HF and LF (High and Low-Frequency components of R-R Interval) for high frequency in the range 0.15–0.4 Hz and 0.04–0.15 Hz for low frequency. Another biomarker that has been used, according to the review, is the EDA, of the selected works, nine have this biomarker in their devices. The GSR sensor can also measure changes in the individual's skin, and some studies show better accuracy when GSR is used. GSR has two main components: Skin Conductance Level (SCL), a slow-changing tonic component, and a faster changing phasic called Skin Conductance Response (SCR). Skin conductance increases when the individual is in a stressful situation. Stress increases the moisture on the skin's surface and significantly alters the individual's flow of electricity. The set of features from the skin changes has other features, as described in [31]. Considering individuals' body temperature and stress level, about five review papers consider skin temperature or body temperature in their devices. And this biomarker is simple to include at the acquisition of a wearable prototype.

6.4 Biomarker Occupational Stress Classifier Model

According to the scientific literature, there is a growing demand regarding the interest in assessing stress to reflect physiological well-being, nutritional status, disease progression, and the immune status compromised by these changes. In healthy individuals, physiological biomarkers have values within a range considered normal (there are acceptable values for individuals considered healthy). However, a stress marker indicates that the individual's body is not in physiological comfort and that different mechanisms that consume energy are operating to maintain homeostasis, significantly altering these values. Biomarkers can be classified as diagnostic, prognostic, and therapeutic biomarkers [7].

The present work presents the study considering biomarkers that can be captured through signals and sensors. There is a stress of euphoria, regarded as positive stress, known as eustress in occupational stress. On occasions when the professional receives a promotion or is satisfied and euphoric with their work activities for achieving goals and objectives, or in situations where the health professional identifies that a treatment given to their patient is showing results, these are positive euphoria regarding work activity. And there is bad or negative stress, called distress, which, if not observed or ignored, can worsen and evolve into burnout [1, 6, 32]. Any stress alters the individuals' homeostasis, altering physiological, physical, and chemical signs in the body due to the presence of stressors, as mentioned above [32]. Based on the review presented in Sect. 2 on biomarkers for identifying stress, features were mapped to classify stress levels in a non-invasive way. This classifier could obtain information from these biomarkers, acquiring signals from wrist wear. A similar classification was presented using ECG and GSR biomarkers, the authors proposed a three levels classification system employing a Finite State Machine and they proved with only two non-invasive signals, the system was able to detect and classify different stress stages [33]. It is essential to emphasize the need to associate information on the professional activity of these health professionals, such as activity, work period, care for COVID patients, access to the ICU, that is, elements of the professional occupation that influence the assessment of the values collected by the device, that is, other stressors related to work activity. Through processing with the help of explainable AI algorithms, it is possible to generate a personalized classification containing different levels of stress for health professionals.

One of the components of a system for HIoT is shown in Fig. 6.2. The HIoT architecture of a classifier for stress levels is based on data from non-invasive sensors (biomarkers) and work context or occupational features. The proposal is a layered architecture to meet the specifics and functionalities of a HIoT system. Among the challenges pointed out is the volume of information collected, transmitted, and processed. The intermediate level (fog or edge computing) is necessary not to overload the network and control the delay of transmission and processing of information due to monitoring the biomarkers. When dealing with data in real-time environments, the delay needs to be minimized so that the information is not discarded due to the deadline of its timestamp. Signal collections could be from a wearable device (imagine a wrist wearable to facilitate day-to-day use). After acquiring this data, they are routed through communication protocols (such as 6LowPAN, or WiFi, or BLE itself, the type of protocol depends on the distance and amount of data that need to be forwarded and the frequency of updating the information).

These issues are being developed through a specific survey for the feasibility of making the wearable. The pre-processing layer performs an analysis and classification of information. Later, it will be transmitted to cloud computing, and from this system, synchronize the results for interested people. One imagines the compilation into a dashboard for the hospital unit's management to monitor the stress levels of its

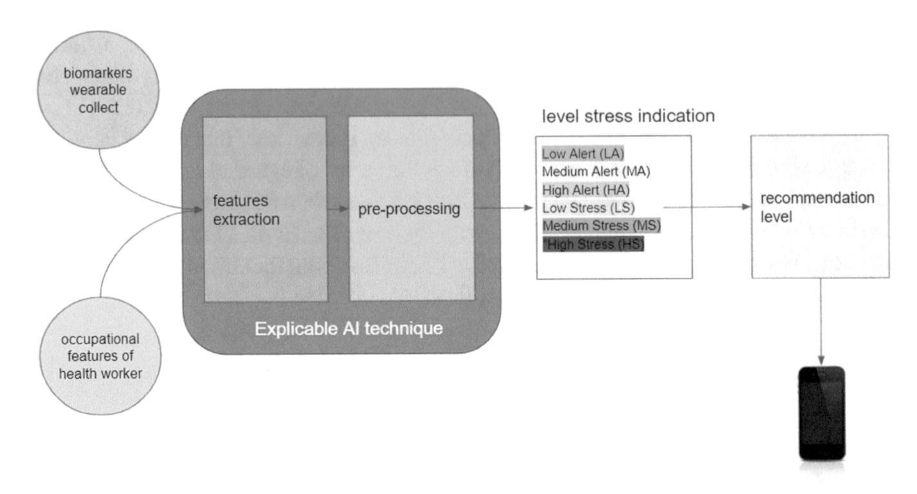

Fig. 6.2 Occupational stress classification for health workers' smart model

team of healthcare professionals. All this effort is part of a research project, in cooperation with the Federal University of Health Sciences of Porto Alegre (UFCSPA) and the Federal University of Santa Catarina (UFSC), with funding from the Research Support Foundation of Rio Grande do Sul (FAPERGS), for the development of an occupational stress surveillance model for health professionals facing the COVID-19 pandemic. It is intended to include a layer with explanations so that the classification is explainable. A layer in the system that can point out the indicators that are pointing to physiological changes and consequently, related to stress classification levels. Figure 6.3 illustrates the layers for H-IoT and the relationship with communication protocols. It is noteworthy that for integration with the platform, the usage of the MQTT (Message Queuing Telemetry Transport) protocol is recommended [28]. The MQTT is employed to transport messages between machine-to-machine IoT platforms and runs over Internet suit protocols like TCP/IP. However, any network protocol that provides ordered, lossless, bi-directional connections can support the MQTT. Considerations regarding the explainable AI will be presented in the next section.

For the discussion here, the following markers should make up individual features need for scaling within the system:

- HRV with low and high frequencies, indicating stress or non-stress levels. This parameter can be used by frequency.
- GSR presence or absence of stress through skin conductance.
- ST (Skin Temperature)—skin temperature, inversely proportional to stress.

Other features will be related to the occupational context, such as type of occupation, weekly hours, rest periods, sleep monitoring, and work in an intensive care unit (ICU). After identifying the algorithm and tests, new information related to labor

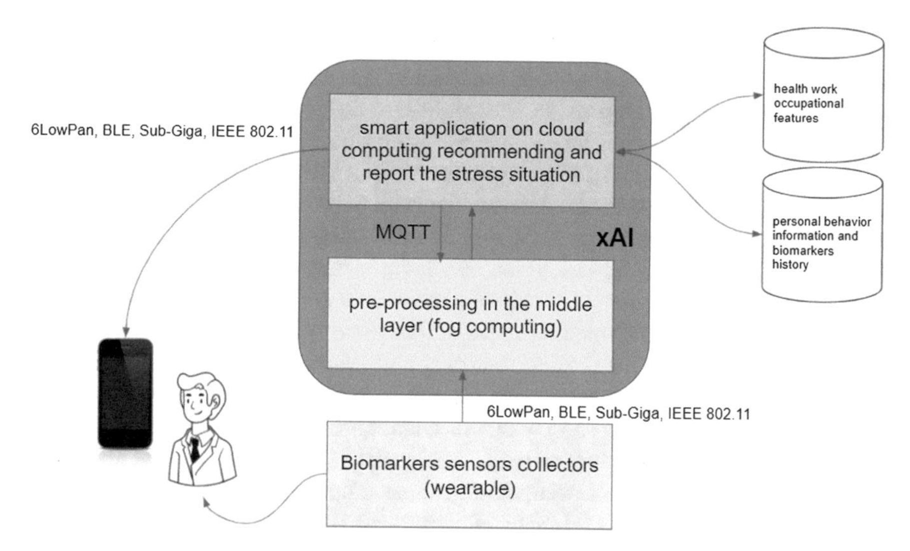

Fig. 6.3 HIoT layers, communication protocols, and functionalities

occupation and that can integrate the set of features for stress classification may be included. This should be done in the next phase of the research project.

6.5 Explainable Artificial Intelligence

Another interesting point is regarding confidence levels. If users cannot trust the model or prediction being made, they will not use it [9], which directly relates to areas of medical decision-making. The question "Would you blindly trust the decision of an AI-based system [34]?" makes us reflect on ethical, social, and scientific issues, which has heightened the interest in machine learning algorithms to be explainable. In recent years, many applications have been based on machine learning algorithms, where extractions and pattern recognition are performed analyzing endless volumes of data, pointing out similarities, statistical analysis, and other functions, often based only on an evaluation accuracy of what has been learned. In these applications, decisions are made based on data analysis of different formats: images, social media, numerical values, text analysis, without considering an ethical analysis of the context in question. The lack of transparency behind their behavior is what leaves users without understanding how these models make decisions. The algorithms are unable to explain the decision-making process of machine learning to human users [35]. Transparency and trust are desirable characteristics when it comes to technological resources, especially in medicine. In complex domain systems, or where decisions involve human lives, such as health, justice, education, and even driving autonomous vehicles, the lack of transparency in models known as black boxes has been severely questioned [34]. To minimize these problems, studies on interpretable machine learning algorithms have advanced. This is part of explainable AI. It is necessary to provide an insight into the logic that AI used to reach a particular conclusion; according to the scientific literature, AI explanations are as old as AI itself [35]. When decisions involve human lives, the context needs to be analyzed, or at least the steps for the decision need to be clear so as not to run the risk of generating a catastrophic situation. Machine learning algorithms, statistical probability algorithms, fuzzy logic, and neural networks have been used to classify work activities in different segments. Analyzing the review articles in Sect. 2, nine studies used the Support Vector Machine (SVM). But most works make use of different machine learning algorithms or deep learning algorithms, all of the "black box" type and without pointing out requirements for interpretability in their proposals, results based on accuracy [11–24]. These methods cannot produce justification for a prediction or classification that they arrive at a new data sample [36]. Other articles in the literature on the study of occupational stress employ algorithms such as SVM, k-Nearest Neighbors (KNN), Linear Discriminant Analysis (LDA), Fuzzy Logic, and Artificial Neural Networks (ANN) [37, 38]. It is noteworthy that these works also present high values related to the accuracy, as in the case of an article that uses only the GSR with 95% accuracy to identify stress using biomarkers of skin conductance change for vehicle driving activities [37]. The authors point out two levels of stress for each segment of the experiment, considering two levels for the low, two for the medium, and two for the high level of stress. The activity of driving vehicles is an activity that, if misunderstood, can generate harmful consequences because it involves people's lives in the case of driving vehicles. Basing decisions solely on the accuracy of the analysis of data collected for just one biomarker is quite risky, even with an accuracy of almost 100%. A model learns from the dataset training with high precision and accuracy. Once learned, previously unseen input can be given to the model, producing a new result without explainable considerations. In this case, about the stress classification, other context features need to be added in the process [36]. Regarding occupational stress, factors such as good (eustress) and bad stress (distress) both present the same physiological changes and demand that the system solution follow some algorithm or mechanism that explains the decision-making process. The information analyzed must include the occupational context of health professionals and not just the physiological signs collected by the wearable sensor. The importance of using a context mechanism in the decision support process is highlighted. Ideally, the system would also include information on various other contexts that change the health conditions of individuals, such as food context, the number of hours of rest, and other factors that imply physiological changes and, consequently, the identification of distress from each individual [7].

6.5.1 Explainability and Interpretability

As already mentioned, the machine learning techniques have no clarity on how the results depend on the input data. The scientific literature has some works about how to explain some machine learning approaches. As the SVM has been more applied in our research about occupational stress and biomarkers techniques, we investigated how to use explanation rules to SVM. A recent paper presents Inductive Logic Programming (ILP) to extract rules to explain SVM machine learning. Induction realizes concept learning through the general sentences from examples and basic knowledge. Representing and reasoning with incomplete knowledge is essential to the ILP technique, as mentioned in [39]. The paper introduces a SHAP-FOIL algorithm that learns a single clause for the most influential support-vector experiments demonstrating the logic program induced by the SHAP-FOIL, which explains a black-box SVM model [40]. It presented a better performance when compared with traditional ILP algorithms. Additionally, the ILP has two approaches, top-down or bottom-up. Each of them has indications for different kinds of problems. For example, top-down is applicable for large-scale or noisy datasets. Meanwhile, the bottom-up is not appropriate on a large scale. The bottom-up approach builds most-specific clauses from the training examples and searches the hypothesis space through generalization usage. On the other hand, the top-down begins with the most general clauses and then specializes them. Through the investigation about how to create explication rules to machine learning, some applicable algorithms for SVM and other machine learning approaches also were encountered [40, 41].

Interpretability also is essential for explainable AI systems. It is a mechanism to provide qualitative understanding between the input variables and the algorithm response [9]. There is no need for a unified version of interpretability. However, [42] claimed there is no formal definition regarding interpretability and presents a significant discussion regarding trust, causality, transparency, and other issues. Furthermore, according to the authors, interpretability should take first, clear reasoning based on falsifiable propositions, and second, offer some natural way of contesting these propositions. And if they are falsified, it must be able to modify the decisions appropriately. There are different types of interpretability to make the algorithms explainable for different kinds of tasks. It will depend on the type of application, the types of data, and if there are mechanisms to associate explanations to the learning process [10]. Of the most varied techniques and algorithms, there are those that process images and need to identify what is in this image, giving interpretability a much greater weight in the context of information processing. On the other hand, some algorithms need to present a classification for a set of numerical samples and their mathematical relationships. Furthermore, some algorithms do not allow the information to be interpretable.

In this context, there are still autonomous systems in which human beings will not analyze decisions. In these cases, they are not systems that involve decisions that change living conditions or put people at risk, and therefore, there is no need to include interpretability [42]. A second category would be systems that can be conditioned to expert decision-making. The system presents its arguments and asks the human to point to the decision (recommendation systems have worked similarly, but there is no transparency in the recommendations). And the system can even be auditable according to its statistical analysis and arithmetic functions. In [34], three other categories are presented for explainable AI (or xAI), opaque systems, those that offer no insight into their algorithmic mechanisms, totally black-boxed; interpretable systems where users can mathematically analyze their algorithmic tools; and comprehensible systems that emit symbols that allow user-guided explanations of how a conclusion is reached. In the scientific literature, there is yet another classification to make machine learning algorithms explainable.

Interpretability can be intrinsic and considered post-hoc, depending on when interpretability is obtained. If it is part of the learning system as a layer, it is considered inherent, and if it is added to explain decisions along the way by pointing out information from the learning process, it is called post-hoc. There is indeed research to help describe the steps in learning classifiers, adding information to understand classification, elucidating black box steps as is the case with Local Interpretable Model-agnostic Explanations (LIME) its variations [34]. The objective of LIME is to identify an interpretable model according to an interpretable representation that is faithful to the classifier. The authors present the main differences between giving features and representations interpretable for human beings in this context. This article also compares the results presented by LIME and other classifiers, such as SVM, random forest, decision trees [9]. Considering the classifier for different types of stress, a way to represent information related to biomarkers, identification of individual values that may indicate a high level of stress or not should be considered. As the problem in question has non-linear characteristics and few features, LIME can be used as a test to interpret the classifier. If these biomarkers register changes and the corresponding relationship with the occupational position, present the individuals' stress level and not just the numerical values. The interpretability here should associate the values collected from the biomarkers and the occupational context of each one. However, the techniques that apply the interpretability and explainability in the stress classification model must be tested, and the outcomes from this investigation work must be considered.

6.5.2 *Explainable Recommendation*

Within the proposed system, there is a layer that is similar to a recommendation system. Recommendation systems emerged in the e-commerce area and advanced to other areas because they present a great way to process a high volume of data and associate recommendation responses to specific user profiles. They can be associated with different techniques, including deep learning to assist in the decision-making process in the medical field [43]. Recommendation systems can make predictions about user consumption. Likewise, there are similar scenarios in the health area where there is an immense amount of data to be processed and analyzed and all the personification of users who work in these environments. As is the case with the medical area and health professionals in a hospital, in this case, it is possible to identify different professionals with different working hours and specific activities. For example, doctors, nurses, nutritionists, physiotherapists, speech therapists, nutritionists, nursing technicians, radiologists, and other people linked to cleaning sectors and other hospital administrative activities. All these professionals may be suffering from different stress conditions due to their work activities due to the Covid-19 pandemic. It is a highly complex problem to measure and assess occupational stress for these different categories automatically.

The recommendations will have to be personalized, and a personalized system is much more complex to develop. Still, stress identification must have individual characteristics and context [44]. However, this system has to be transparent. Recommendations need to be explainable. The recommendation system will be able to employ deep learning techniques to identify these different categories in processing information related to stress and how to point out to users how to face the situation so that it does not become a more serious situation and a case of seeking help from professionals. In the scientific literature there are articles relating interpretability to deep learning techniques [10, 35, 45, 46]. Several models implement explainable recommender systems such as a relevant user or item explanation, feature-based, opinion-based, sentences, visual or social explanation [47, 48]. A biomarkers and occupational stress recommender system proposed in this work should aggregate medical or psychologists' collective opinions in expert-generated content as explanations and sentence and visual explanations. Still, for each set of professionals and individual features, a diversified series can be recommended, respecting the individuality of each health professional. Each of these professionals has a specific workday. In addition to the workload related to the hours dedicated to the intensity of their work activity, there are other factors considered stressors, such as whether they work closer to the treatment sectors of patients with Covid-19, whether they attend ICUs, whether they have access and contact with patients with Covid-19. These factors can be considered stressors and should be included in a system that analyzes physiological signs and this information from the work context. The importance of context for a decision-making process and the relevance of interpretability are highlighted. The recommendation needs to be explainable, and it is essential to have interpretability by users.

Otherwise, there is a risk of not having enough confidence in the recommendations. The ability to explain how the algorithm works or how the proposal relates to the decision-making process is essential for implementing this type of solution in a hospital network. Another essential feature is feedback. Based on the use of the system, a specific recommendation is evaluated as positive or negative, based on the new data collected by the sensors. Observing the behavior of biomarkers makes the system intelligent and adaptable to the responses of its users. The system can improve its recommendations based on user responses regarding the activities performed. For example, the system recommends the practice of meditating or walking according to the identified stress level. The system records the results of biomarkers from the activity chosen by its user, customizing the system for individual use. Figure 6.4

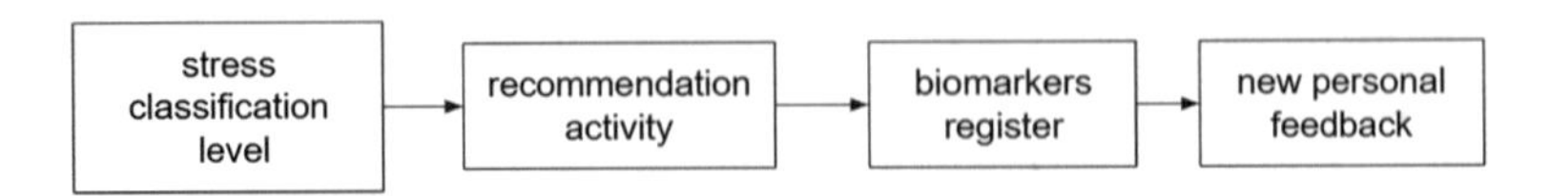

Fig. 6.4 Feedback scheme to record new biomarker values from a recommendation activity

presents the flow for the feedback functionality to adjust the personal response about the recommendation into the system if it was a good or bad response.

All this information is related to a system for recommending activities that can minimize the identified stress conditions. For example, by identifying lower stress levels, the system may recommend actions related to meditation and relaxation activities or outdoor physical activities. In the case of identification of distress symptoms, the help of a qualified professional to conduct a more specific treatment should be recommended (in this case, the system may associate physicians who are eligible to act as qualified professionals for this issue). This system, in the future, in addition to helping professionals deal with stress, should be able to map the mental health conditions related to occupational stress of health professionals working on the front lines of the pandemic or other infectious, contagious diseases that arise. It is a tool designed to support managers of health units to help identify the conditions of the professional team and support the definition of the team's work schedule.

6.6 Data Security and Privacy

Data security and privacy are still a gap to be addressed for H-IoT systems and devices. In the scientific literature, several works investigate techniques and algorithms to solve this problem. It is possible to project some questions for the model proposed in this chapter through some specific points. As with any networked device, there is a potential risk due to its vulnerability; stress classification devices should also be able to protect themselves through techniques and algorithms that have been recommended for different levels of the proposed architecture. Remember that devices in the health area (healthcare) need special care because they can put the lives of human beings at risk. In this case, it is no different. We do not want to expose private information about the physical and mental conditions of health professionals. Considering the layers provided throughout the text, it is possible to identify points that need attention and techniques to be investigated to meet the security and privacy of information. In the case of devices and BLE use, care must be taken with the help of protocol for discovering plug and play services or protocol properties used to direct IoT devices [27]. The literature points out device authentication mechanisms associated with cryptography as an alternative to simpler devices, and these can be adapted to the model discussed in this chapter. To prevent attacks, it is possible to block unauthenticated requests at the device level. Furthermore, due to the limited processing and computational power of IoT devices, the use of cryptography with symmetric algorithms is recommended.

These algorithms use codes for message and network authentication, usually increasing the communication load, the need for storage space, and consequently having higher energy consumption. However, they are still the only ones that can be processed in devices used in IoT, that is, devices with low computational power [49]. Regarding the network or communication of devices, protocols that have layers of routing and security are recommended. The WIFI itself allows adding encryption at the network level. But it is also possible to improve the solution by incorporating hashing and MD5 mechanisms into the messages. However, these mechanisms will increase the communication load, and it needs to be investigated if this procedure is adequate to identify the biomarkers predicted in the model. When it comes to recording information in real-time, the added time due to these mechanisms can cause an unwanted delay to the system [12].

There is still a concern with intermediate and cloud levels, recommending the use of services concerned with data security and privacy. It is possible to establish a set of strategies to combat the negative impacts generated by cyber-attacks proactively. To end this section, there is a concern with the human factor. Health professionals who are going to use the devices must be trained to understand the importance of monitoring information on an ongoing basis and make sure that the devices will be used in a personal and non-transferable way for the success of the personalized recommendations. Possibly, after training, it is necessary to sign a term of commitment, which can be stored in the cloud system itself, certifying that the person is aware of their obligations regarding using this device. There is still a need to respect the privacy of the information collected and analyzed by the system to avoid public exposure of the health professionals involved in the research. Device sharing, access keys, use of weak passwords, dissemination of results should be considered in this process concerning human factors involved in the project.

6.7 Conclusion

This chapter addressed challenges related to developing a model for identifying occupational stress in health professionals through non-invasive biomarkers and intelligible AI resources. The development of a system that allows the recommendation of actions according to the level of stress that is intelligible, transparent, and that, through these resources, is trustworthy for health professionals to adopt this system. There are several challenges in implementing a proposal at this level, such as setting a stress classification at different custom levels and developing all HIoT system layers and their communication, processing, and information synchronization processes. Moreover, the challenge of dealing with information security and privacy in the layers foreseen for the HIoT system. Additionally, several studies about machine learning and biomarkers usage are related. These investigations represent an effort to employ better the technology connecting sensors and IoT to create new resources to caption information and collaborate with health professionals in this area.

Acknowledgements Foundation of Rio Grande do Sul (FAPERGS) for supporting research 20/2551–0000262-4.

References

1. Martins, L.F., Laport, T.J., Menezes, V.P., Medeiros, P.B., Ronzani, T.M.: Burnout syndrome in primary health care professionals. Esgotamento entre profissionais da Atenção Primária à Saúde. Ciência & Saúde Coletiva (online), v. 19, n. 12, pp. 4739–4750 (2014). https://doi.org/10.1590/1413-812320141912.03202013
2. Stacciarini, J.M., Tróccoli, B.T.: Instrumento para mensurar o estresse ocupacional: inventário de estresse em enfermeiros (IEE). Rev. Latino Am. de Enfermagem **8**(6), 40–49 (2000). https://doi.org/10.1590/s0104-11692000000600007
3. Paschoal, T., Tamayo, A.: Validation of the work stress scale. Validação da escala de estresse no trabalho. Estudos de Psicologia (Natal) (online) v. 9, n. 1, pp. 45–52 (2004). https://doi.org/10.1590/S1413-294X2004000100006
4. Koh, D.: Occupational risks for COVID-19 infection. Occup. Med. **70**(1), 3–5 (2020). https://doi.org/10.1093/occmed/kqaa036
5. Coghi, M.F., Coghi, P.F.: Stress e ansiedade: eles estão te consumindo? In: Anais do 14o Congresso de Stress do ISMA BR, no 14 (2013)
6. Moayed, M.S., et al.: Survey of immediate psychological distress levels among healthcare workers in the COVID-19 epidemic: a cross-sectional study. In: Clinical, Biological, and Molecular Aspects of COVID-19, vol. 1321, P. C. Guest, Org., pp. 237–243. Springer International Publishing, Cham (2021). https://doi.org/10.1007/978-3-030-59261-5_20
7. Dhama, K., et al.: Biomarkers in Stress-Related Diseases/Disorders: Diagnostic, Prognostic, and Therapeutic Values. Front. Mol. Biosci. **6**(91), out. (2019). https://doi.org/10.3389/fmolb.2019.00091
8. Guidotti, R., Monreale, A., Ruggieri, S., Turini, F., Pedreschi, D., Giannotti, F.: A survey of methods for explaining black-box models, vol. 51, no. 5. arXiv (2018)
9. Ribeiro, M.T., Singh, S., Guestrin, C.: Why should I trust you?: Explaining the predictions of any classifier. In: Proceedings of the 22nd ACM SIGKDD International Conference on Knowledge Discovery and Data Mining, pp. 1135–1144. San Francisco California USA, ago (2016). https://doi.org/10.1145/2939672.2939778
10. Tjoa, E., Guan, C.: A survey on explainable artificial intelligence (XAI): towards medical XAI. IEEE Trans. Neural Netw. Learn. Syst. 1–21 (2020). https://doi.org/10.1109/TNNLS.2020.3027314
11. Moher, D., Liberati, A., Tetzlaff, J., Altman, D.G., PRISMA Group.: Preferred reporting items for systematic reviews and meta-analyses: the PRISMA statement. Ann. Intern. Med. **151**(4), 264–269, W64, ago. (2009). https://doi.org/10.7326/0003-4819-151-4-200908180-00135
12. Carreiro, S., Chintha, K.K., Shrestha, S., Chapman, B., Smelson, D., Indic, P.: Wearable sensor-based detection of stress and craving in patients during treatment for substance use disorder: A mixed-methods pilot study. Drug Alcohol Depend. **209**, 107929 (2020). https://doi.org/10.1016/j.drugalcdep.2020.107929
13. Kumar, A., Sharma, K., Sharma, A.: Hierarchical deep neural network for mental stress state detection using IoT based biomarkers. Pattern Recognit. Lett. **145**, 81–87 (2021). https://doi.org/10.1016/j.patrec.2021.01.030
14. Patlar Akbulut, F., Ikitimur, B., Akan, A.: Wearable sensor-based evaluation of psychosocial stress in patients with metabolic syndrome. Artif. Intell. Med. **104**, 101824 (2020). https://doi.org/10.1016/j.artmed.2020.101824

15. Izumi, K., et al.: Unobtrusive sensing technology for quantifying stress and well-being using pulse, speech, body motion, and electrodermal data in a workplace setting: study concept and design. Front. Psychiatry **12**: 611243 (2021). https://doi.org/10.3389/fpsyt.2021.611243
16. Montesinos, V., Dell'Agnola, F., Valdés, A., Aminifar, A., Atienza, D.: Multi-Modal Acute Stress Recognition Using Off-the-Shelf Wearable Devices, vol. 2019 (2019). https://doi.org/10.1109/EMBC.2019.8857130
17. Dalmeida, K.M., Masala, G.L.: HRV features as viable physiological markers for stress detection using wearable devices. Sensors **21**(8), 2873 (2021). https://doi.org/10.3390/s21082873
18. Clark, J., Nath, R.K., Thapliyal, H.: Machine Learning-Based Prediction of Future Stress Events in a Driving Scenario. ArXiv210607542 Cs Eess (2021)
19. Sánchez-Reolid, R., Martínez-Rodrigo, A., López, M.T., Fernández-Caballero, A.: Deep support vector machines for the identification of stress condition from electrodermal activity. Int. J. Neural Syst. **30**(7), 2050031 (2020). https://doi.org/10.1142/S0129065720500318
20. Kaczor, E.E., Carreiro, S., Stapp, J., Chapman, B., Indic, P.: Objective measurement of physician stress in the emergency department using a wearable sensor. In: Proceedings of Annual Hawaii International Conference System Science, vol. 2020, pp. 3729–3738 (2020)
21. Betti, S., et al.: Evaluation of an integrated system of wearable physiological sensors for stress monitoring in working environments by using biological markers. IEEE Trans. Biomed. Eng. **65**(8), 1748–1758 (2018). https://doi.org/10.1109/TBME.2017.2764507
22. Dzieżyc, M., Gjoreski, M., Kazienko, P., Saganowski, S., Gams, M.: Can we ditch feature engineering? End-to-end deep learning for affect recognition from physiological sensor data. Sensors **20**(22), 6535 (2020). https://doi.org/10.3390/s20226535
23. Subhani, A.R., Mumtaz, W., Saad, M.N.B.M., Kamel, N., Malik, A.S.: Machine Learning Framework for the Detection of Mental Stress at Multiple Levels, pp. 13545–13556. IEEE Access, vol. 5 (2017). https://doi.org/10.1109/ACCESS.2017.2723622
24. Han, L., Zhang, Q., Chen, X., Zhan, Q., Yang, T., Zhao, Z.: Detecting work-related stress with a wearable device. Comput. Ind. **90**, 42–49 (2017). https://doi.org/10.1016/j.compind.2017.05.004
25. Höller, J. Org.: From Machine-to-Machine to the Internet of Things: Introduction to a New Age of Intelligence. Elsevier Academic Press, Amsterdam (2014)
26. Morales, A.S., de O. Ourique, F., Cazella, S.C.: A Comprehensive review on the challenges for intelligent systems related with internet of things for medical decision. In: Marques, G., Kumar Bhoi, A., de la Torre Díez, I., Garcia-Zapirain, B. (eds.) Enhanced Telemedicine and e-Health: Advanced IoT Enabled Soft Computing Framework, pp. 221–240. Springer International Publishing, Cham (2021). https://doi.org/10.1007/978-3-030-70111-6_11
27. Firouzi, F., Farahani, B., Ibrahim, M., Chakrabarty, K.: Keynote paper: from EDA to IoT eHealth: promises, challenges, and solutions. IEEE Trans. Comput.-Aided Des. Integr. Circuits Syst. **37**(12), 2965–2978 (2018). https://doi.org/10.1109/TCAD.2018.2801227
28. Qadri, Y.A., Nauman, A., Zikria, Y.B., Vasilakos, A.V., Kim, S.W.: The future of healthcare internet of things: a survey of emerging technologies. IEEE Commun. Surv. Tutor. **22**(2), 1121–1167 (2020). https://doi.org/10.1109/COMST.2020.2973314
29. Greene, S., Thapliyal, H., Caban-Holt, A.: A survey of affective computing for stress detection: evaluating technologies in stress detection for better health. IEEE Consum. Electron. Mag. **5**(4), 44–56 (2016). https://doi.org/10.1109/MCE.2016.2590178
30. Zanon, V.R., Romancicni, E.M.R., de O. Ourique, F., Morales, A.S.: Dispositivo com Interface Vestível para a Aquisição, Processamento e Transmissão do Sinal Cardíaco em Exame de Eletrocardiograma. In: Anais do Simpósio Brasileiro de Computação Aplicada à Saúde (SBCAS), pp. 48–59 (2021)
31. Nath, R.K., Thapliyal, H., Caban-Holt, A., Mohanty, S.P.: Machine learning-based solutions for real-time stress monitoring. In: IEEE Consumer Electronics Magazine, vol. 9, no. 5, pp. 34–41, 1 (2020). https://doi.org/10.1109/MCE.2020.2993427
32. Sharma, N., Gedeon, T.: Objective measures, sensors and computational techniques for stress recognition and classification: a survey. Comput. Methods Programs Biomed. **108**(3), 1287–1301 (2012). https://doi.org/10.1016/j.cmpb.2012.07.003

33. Martinez, R., Irigoyen, E., Arruti, A., Martin, J.I., Muguerza, J.: A real-time stress classification system based on arousal analysis of the nervous system by an F-state machine. Comput. Methods Progr. Biomed. **148**, 81–90 (2017). https://doi.org/10.1016/j.cmpb.2017.06.010
34. Doran, D., Schulz, S., Besold, T.R.: What Does Explainable AI Really Mean? A New Conceptualization of Perspectives. ArXiv171000794 Cs (2017) http://arxiv.org/abs/1710.00794
35. Du, M., Liu, N., Hu, X.: Techniques for interpretable machine learning. Commun. ACM **63**(1), 68–77 (2019). https://doi.org/10.1145/3359786
36. Lipton, Z.C.: The mythos of model interpretability. arXiv e-prints (2016)
37. Memar, M., Mokaribolhassan, A.: Stress level classification using statistical analysis of skin conductance signal while driving. SN Appl. Sci. **3**(1), 64 (2021). https://doi.org/10.1007/s42452-020-04134-7
38. Gul Airij, A., Bakhteri, R., Khalil-Hani, M.: Smart wearable stress monitoring device for autistic children. J. Teknol. **78**, 7–5 (2016). https://doi.org/10.11113/jt.v78.9453
39. Sakama, C.: Induction from answer sets in nonmonotonic logic programs. ACM Trans. Comput. Log. **6**(2), 203–231 (2005). https://doi.org/10.1145/1055686.1055687
40. Shakerin, F., Gupta, G.: White-box induction from SVM models: explainable AI with logic programming. ArXiv200803301 Cs (2020)
41. Diederich, J.: Rule Extraction from Support Vector Machines, Studies in Computational Intelligence, vol. 80. Springer International Publishing. ISSN 1860-949X (2008)
42. Holzinger, A., Biemann, C., Pattichis, C.S., Kell, D.B.: What do we need to build explainable AI systems for the medical domain? ArXiv171209923 Cs Stat (2017)
43. Batmaz, Z., Yurekli, A., Bilge, A., Kaleli, C.: A review on deep learning for recommender systems: challenges and remedies. Artif. Intell. Rev. **52**(1), 1–37 (2019). https://doi.org/10.1007/s10462-018-9654-y
44. Zhang, S., Yao, L., Sun, A., Tay, Y.: Deep learning-based recommender system: a survey and new perspectives. ACM Comput. Surv. **52**(1), 1–38 (2019). https://doi.org/10.1145/3285029
45. Guidotti, R., Monreale, A., Ruggieri, S., Turini, F., Giannotti, F., Pedreschi, D.: A survey of methods for explaining black-box models. ACM Comput. Surv. **51**(5), 1–42 (2019). https://doi.org/10.1145/3236009
46. Guidotti, R., Monreale, A., Ruggieri, S., Pedreschi, D., Turini, F., Giannotti, F.: Local Rule-Based Explanations of Black Box Decision Systems. ArXiv180510820 Cs (2018)
47. Abdollahi, B., Nasraoui, O.: Transparency in fair machine learning: the case of explainable recommender systems. In: Zhou, J., Chen, F. (eds.), Human and Machine Learning, pp. 21–35. Springer International Publishing, Cham (2018). https://doi.org/10.1007/978-3-319-90403-0_2
48. Zhang, Y., Chen, X.: Explainable recommendation: a survey and new perspectives. Found. Trends® Inf. Retr. **14**(1), 1–101 (2020). https://doi.org/10.1561/1500000066
49. Jing, Q., Vasilakos, A.V., Wan, J., Lu, J., Qiu, D.: Security of the Internet of Things: perspectives and challenges. Wirel. Netw. **20**(8), 2481–2501 (2014). https://doi.org/10.1007/s11276-014-0761-7

Part II
Smart Cities

Chapter 7
Smart Cities, The Internet of Things, and Corporate Social Responsibility

Andrew D. Roberts

Abstract By 2050, approximately 66% of the world's population will live in urban centres. Economic, social and environmental transformation, and inherent challenges are expected. Creating smarter cities through the use of Information and Communication Technology (ICT) and capitalising on Internet of Things (IoT) expertise, are expected to resolve many of these challenges. Over the last decade, a plethora of studies has been undertaken regarding the use of these technology(ies) in diverse areas. Findings indicate operational efficiencies, improvements in infrastructure service(s), and environmental and societal benefits. However smart city implementations create inadvertent challenges, many arising from IoT usage itself. This chapter begins by exploring how smart ICT and IoT devices can be used to resolve urbanisation problems, including identifying the inherent risks, problems, issues and challenges with IoT use. Building on Corporate Social Responsibility (CSR) literature, this chapter argues that policymakers must re-orientate stakeholder engagement in the IoT's decision-making process. Proposing that, through meaningful and productive stakeholder collaboration, via simultaneous top-down and bottom-up evolving engagement processes, relevant IoT risks, problems and issues can be addressed.

Keywords Smart City(ies) · ICT · IoT · CSR · CSV · Governance

7.1 Introduction

By 2050, researchers suggest that the percentage of the world's population living in urban centres will increase from 50 to 66% [1, p. 77], whereby increased productivity, social and human capital growth and knowledge sharing, idea creation and innovation remain an inevitable outcome. However, poorly managed urban growth can lead to urban sprawl and housing affordability issues; increased poverty; unemployment and urban costs; rising inequality; environmental degradation [2, 3]; and poor citizen health and wellbeing [4]. Commentators suggest that growing economic, population

A. D. Roberts (✉)
Central Queensland University, Queensland, Australia
e-mail: a.d.roberts@cqu.edu.au

© The Author(s), under exclusive license to Springer Nature Switzerland AG 2022

G. Marques et al. (eds.), *Machine Learning for Smart Environments/Cities*,
Intelligent Systems Reference Library 121,
https://doi.org/10.1007/978-3-030-97516-6_7

and suburban spread leads to an exponential rise in: urban waste, by 2025 global urban waste is predicted to exceed 2.2 billion tonnes [5]; and threats to water sustainability, due to water loss from poorly managed water supply systems [6], changes in ground water levels, exhaustion of local water resources, water pollution and drought [7]. As a continuum, it is further suggested that as cities grow: passenger vehicle usage will increase, whereby global numbers are expected to reach 2.6 billion by 2050, resulting in increased traffic congestion and air pollution [8]; emerging challenges with electricity reliability and supply [9]; and a growing gap between healthcare availability and accessibility, exacerbated by a disproportionate allocation of qualified medical personnel across different geographical locations [10].

It is suggested that addressing such challenges requires the use of Information and Communication Technology (ICT) to make our cities smarter (i.e., by creating smart city(ies)), and transform how they economically, environmentally, and socially operate [2, 11]. Smart cities enable municipal governments to create scalable, self-ruling, self-managing, self-sustainable systems that re-configure and re-integrate communities [12]. Simultaneously smart cities address social, economic, and environmental issues, improve urban services and operational efficiency [13–15], improve societal wellbeing [16, 17], and create safer and culturally vibrant communities [18, 19]. Consequentially, municipal governments have been looking at ways to use ICT, in particular the Internet of Thing's (IoT's), to address these challenges. IoT is defined as "a physical or virtual object which connects to the Internet and has the ability to communicate with human users or other objects" [20, p. 230], that "integrate[s] a system to form a network of communication with beings or things" [21, pp. 39–40]. Relying on wireless and static network communications and advanced computing capabilities, IoT devices can smartly connect artefacts, sensors, cameras, utility services, control management systems, transport monitoring systems and other objects [22, 23]. IoT devices enable real-time data production, analysis, and aggregation monitoring, to support decision making and control optimal resource usage to regulate, control, manage and automate how a city function(s) [24].

However, the use of IoT's and ICT triggers significant changes and challenges to organisations and society through an inevitable evolving process that has impacts on and implications for Corporate Social Responsibility (CSR), performance and the environment [2, 12]. Literature indicates that such change does not necessarily generate positive outcomes for all. Inadvertently, expected change can make it more difficult for citizens from low socioeconomic, ethnic, and different religious backgrounds to integrate with their communities [25, 26]. This chapter suggests that to some degree these issues have emerged by somewhat excluding businesses from the process [27] and leaving the decisions on the problems to be addressed, including how best to use ICT, including IoT's, to city governments alone [28, 29]. Yet, business(es) has a vital role to play in resolving societal and environmental issues within our cities [2]. By leveraging the CSR literature, this chapter will make a case for a more inclusive, meaningful, and productive collaboration between policymakers, urban planners, and business(es) in the decision-making process, to maximise the benefits and address some of the challenges of IoT use.

The chapter begins by exploring the contribution(s) business makes to smart city ecosystems. Questions are raised concerning some group(s) in society who remain excluded from, and or have limited opportunity(ies) to benefit from the implementation(s) of ICT and IoT within smart cities. A case is made for a re-conceptualisation of how municipal governments work with business(es) and citizens to make cities smarter, and more environmentally and socially sustainable. IoT applications in smart cities, the benefits, the problems they address, and their challenges are discussed. The chapter concludes by proposing a model that enhances collaboration between businesses and municipal governments that is more inclusive, meaningful, and productive, to resolve the challenges, problems and issues identified, and to improve economic, social and environmental sustainability.

7.2 Smart Cities, Business and CSR

It has been questioned the potential for ICT technology, including IoT's, to simultaneously address the problems arising from urbanisation, and further, enhance economic and social health and wellbeing [2]. Questions have been raised about one-sided polarised ICT perspective(s) on smart city design and the way in which the smart city(ies) agenda is driven by large technology vendors and consultancies alone (e.g., IBM, Google, CISCO, and PwC) [2]. In Ref. [30], the author raises issues about the overemphasis on technology as a solution to all ills. Suggesting that doing so limits how policymakers develop smart city strategies. Such a myopic and biased focus ignoring considerations of other important factors including citizens social, liveability and mobility needs, and environmental factors and governance. Others have argued, that maintaining an emphasis on technology, limits sustainable social development by failing to consider basic social needs [31]. And, that it fails to recognise the complex and multi-dimensional nature of cities [2]. That cities are "complex systems of systems" [2, p. 3], adapting to changing environments through "different space–time patterns of nodes and links" [32, p. 8]. Adaptations that asynchronously have an impact on, and are impacted by, multiple stakeholders and groups, including businesses, communities, and citizens, with diverse, conflicting, and competing needs, expectations and demands [33, 34].

Questions have also been raised about the role played by large technology vendors and corporations, such as IBM, in the drive to design and develop a smart city direction and goals. It is argued that this top-level large corporate involvement inadvertently creates an unnecessary sense of competition between these vendors and city governments [2, 35]. In that, large technology vendors, rather than city governments, drive the decisions and actions needed to develop and implement smart city(ies) strategies. Thus, creating unnecessary tension with policymakers for ownership, authorship, the authority of, and the profit arising from a smart city(ies) implementation(s). Roberts [2] explores these issues, using IBM as an example, by questioning IBM's view, that they alone can provide expert solutions to address urban problems, through their intelligent operations centre. Arguing that this perspective redefines the function of city administrator(s) by assuming a level of homogeneity for all city(ies), that one-size-fits-all, that in reality, does not exist. Moreover, in Ref. [2] the author argues the IBM smart city solution(s) creates several problems. Firstly, IBM assumes they can, reform and improve how a city works from the top-down, and somehow shape it into reality through language and rhetoric. Secondly, that they can implement individual solutions to address each urban problem (e.g., water management or smart buildings). Thirdly, that city infrastructure(s) already exist. And lastly, that cities are homogenous with common administration function(s) [2, 36].

However, cities are non-homogenous complex systems of systems, made up of many overlapping worlds, that in many cases are incommensurable, and operate with differing administrative functions [2, 35]. Such cities, cannot be reformed and or re-designed from the top-down, shaped into a reality only driven and known by large corporates such as IBM. Indeed, in many cities, infrastructure is non-existent, broken or worsening [2, p. 5, 36]. Responding to urban problems, using technology only, demands more than language and rhetoric. Individual, specific homogenous approaches, such as IBM's singular technology approach, fails to recognise the overlapping and complex nature of environmental and societal problems (e.g., urban transport congestion impacting air and environmental pollution) [2]. Rather, interconnected, and integrated systems are needed to facilitate and support policymaker(s) action taking and decision making through and across, system data and information collection, aggregation, and analysis.

IBM's approach to smart cities may in addition, intentionally or unintentionally, exclude(s) non-mainstream, disenfranchised and somewhat powerless stakeholder groups and citizens. Potentially considered irrelevant, disturbances and irritations with their concerns, such people groups could easily be ignored by such big corporations [2, 35]. Further, IBM's ultimate business vision fails to recognise the way in which smart city technology shapes and re-shapes citizens behaviour, or to take into consideration associated, positive and negative externalities [2]. For example, smart city technology can both enhance and grow citizen(s) digital literacy [37], and or leave citizens feeling vulnerable and under surveillance, subject to their perception of the way in which interconnected big data is collected and used [38]. Further, it could be argued that by failing to recognise the complex inter-connecting web of socio-cultural and economic structures and processes, deploying IBM's smart city technology alone, may create social exclusion, by leaving little space for disadvantaged groups [31] and or transient populations [39]. Roberts [2] argues, that in line with potential big corporation 'take-over' and influence in such complex evolving city functioning and living health, citizens rightly require and are demanding, inclusion in smart city design and ultimate decision-making processes.

Commentators have further criticised the lack of consideration of the interests and practical impacts and implications for and on citizens of policymakers in the smart city design, decision making and implementation process. Kitchin [37] argues, citizen interests and views seem to be somewhat ignored and subjugated by smart city policymakers. Consequentially, smart cities are in danger of shifting citizenship "away from civic responsibility and engagements" to an economic exchange enacted through data sharing and "classifying [citizens] as consumers who purchase services"

[40, p. 15]. Thus, questions are emerging about the degree to which policymakers must engage with citizens in the identification of social and environmental problems and their solution(s) [2]. To address these issues, scholars have argued that smart cities need to be more citizen-focused [41], and that the smart city vision must be translated into and accommodate for the political, economic, and social context of a city's citizens [2, 36]. This requires identifying the needs of citizens, their perspective on the problems and issues in their communities, and the implementation of the relevant and practical solution(s).

Others have argued that urban planners need to: search for socio-techno-synergies in urban systems; recognise that aligning social and cultural structures with technology is not a simple process; alter how they view a smart city; and, accept and address the complexity associated with institutional change that implementing a smart city design presents [42]. Addressing these problems and issues require recognising and engaging with multiple stakeholders in smart city design, and implementation process(es), including business(es), community groups and citizens, at every stage of a smart city project [2, 43]. This can be achieved through active stakeholder engagement, by working with business(es) to leverage their willingness to engage in CSR to Create Shared Value (CSV) [44–46], while recognising and embracing the valuable contribution they can make to the identification and resolution of environmental and societal issues and concerns [2, 47].

7.3 IoT's—The Benefits

By 2025, it is predicted that 100 billion integrated IoT devices will be deployed in urban domains [48, 49], to organise, manage and control urban resource(s) and infrastructure(s) [11, 16, 50]. Used appropriately, IoT device(s) can monitor, manage, control, and improve: waste collection and disposal [51, 52]; water supply management [53]; air pollution by providing information on air quality violations [54]; electricity supply reliability for consumers and industry [9, 55]; control and management of urban transport system(s) [56]; and, improve healthcare access and affordability for all citizens through the creation of smart health systems [10]. Such predictions have led to a plethora of pilot studies of IoT usage within city ecosystems, which when deployed to scale, have the potential to make our cities more environmentally sustainable [50, 57].

7.3.1 IoT's: Waste, Water Supply and Air Pollution

Given that 80% of waste management costs arise from garbage trucks within our cities as they collect household waste, research efforts on the use of IoT's have been focused on optimising waste collection [52]. For example, researchers have combined ultrasonic IoT sensors, microcontrollers, and Supervisory Control and Data Acquisition (SCADA) technology to pilot end-to-end waste management systems. These ultrasonic IoT sensors identify and send real-time data concerning the volume and type of waste per bin and their geo-location(s). Such smart data facilitates smart sorting of waste into bio-degradable materials for use in localised bio-generation plants and non-biodegradable alternatives. With such smart changes, findings indicate a reduction in greenhouse gas emissions, in addition to improved garbage truck route optimisation and more efficient waste collection [58].

However, efficient waste collection practices form only part of the picture in urban settings. Raising citizen awareness and encouraging good citizenship behaviours (e.g., giving them access to information on where and how to effectively dispose of household waste), including recycling, to balance collection schedules against demand is also required [52]. In reference [59], findings from their pilot smart-waste system further reinforce such viewpoints, whereby through the use of IoT sensors, open-source technology(ies) and a digital application to collect and provide citizens with real-time data on waste volume(s) per garbage bin, their geolocation, and waste collection schedules, increasing citizens awareness and reinforcing good citizen practices of where, how, and when to dispose of household waste further supported and strengthened municipal recycling efforts.

IoT sensors have also been used in the water industry for leakage and spill detection. Researchers [60] found that combining IoT sensors with software that performs algorithmic analysis, was superior to visible evaluation by experienced personnel who literally walked along pipelines to look for anomalies. Consequentially, several pilot studies have been undertaken using various combinations of different IoT sensors and software technologies to monitor and address water leakage. For example, researchers deployed sensors placed along multiple flow hydrant points, to virtually collect data for the measurement and detection of water leakage in and across a distributed reservoir water supply system. Findings indicated a reduction in leakage monitoring costs and more efficient hydrant maintenance [60, 61].

In the US, schematic(s) for IoT and big data-based system(s) have also been developed to manage the national water supply system that supply's water to several hundred million consumers through more than 170,000 water utility companies. The schema connects downstream and upstream data on water flow and usage from an IoT wireless sensor network and an automated water reading system. Findings highlighted improvement considerations in system performance and efficiencies in identifying water fraud and loss [62]. IoT devices for water supply management are also being deployed at scale. For example, in Australia, Telstra (a telecommunications provider) and Yarra Valley Water (YVW) are deploying 1 million IoT devices across YVW water network. In partnership, Telstra will provide the IoT devices, connectivity, and big data software, enabling YVW to collect and analyse real-time data across the water supply system. Such data is expected to provide a greater understanding of the water management infrastructure, proactive leakages, and spills management, further informing improved services and lower costs for consumers [63].

Pilot studies have also been undertaken with IoT devices on air pollution detection and reporting. For example, various municipal regulators have deployed IoT devices and microcontrollers in industrial areas to collect and analyse air pollution and the concentration of different atmospheric gas pollutants at various industrial sites and provide real-time measurement, management, and control of air pollution. Findings from these studies are positive. Continuous monitoring, recording, and analysing of the variability of air pollution and air quality data at different geo-locations, has enabled early intervention and more precise decision making to inform and support the implementation of specific air pollution control mechanisms by government agencies [64, 65].

Researchers have also piloted and validated IoT based air pollution monitoring and control systems for underground rail networks. With study outcomes accurately predicting air pollution levels, available data further enables and informs more accurate location of air purifiers to improve citizens health and wellbeing [66]. Other studies have reportedly attached geomatic IoT sensors to bicycles, with additional mobile devices added as required, to pilot mobile air pollution monitoring system(s). Such studies endeavour to collect and share air pollutant data, and pollutant densities across different city districts, between citizens and municipal governments. Findings suggest that involving citizens in pollution monitoring enhances citizenship, increases their awareness of pollution levels at different localities, and improves overall pollution management [67].

7.3.2 IoT's: Urban Transportation and Electricity Supply

For some years, researchers have explored how IoT devices can improve urban transport networks using predictive routing algorithms, to predict and optimise traffic networks for road users. Real-time data from these devices is aggregated to monitor road traffic and control and manage traffic management systems. Findings from these studies repeatedly demonstrate an improvement in traffic flow and congestion reduction [68]. Ant colony optimisation methods and IoT devices have been used to simulate a distributed intelligent traffic system. Using traffic speed as an indicator of traffic levels, predictive algorithms were shown to efficiently compare alternative routing scenarios and to re-route traffic to less congested roads [69]. Pilot studies have also been undertaken using IoT devices and group algorithms to correlate the impacts of driving behaviours, patterns, and habits. Real-time data collected and used to route and re-route vehicles, based on these characteristics. Initial findings indicated a reduction in traffic congestion and a lowering of motor vehicle dependency from citizens [70].

In the electricity sector, researchers [71] developed a scalable, low-cost, flexible, and secure experimental home automation system using IoT's to connect customers cell phones to home electricity systems, enabling remote control and management of their electricity use. IoT smart systems have also been deployed to connect buildings to smart servers, enabling consumers to centrally automate, manage and control electricity consumption across multiple locations [72]. With a view to deploying IoT solutions at scale to manage electricity supply, researchers have developed frameworks that improve the integration of electrical devices and systems to increase the interoperability and data communication between industrial equipment [73, 74]. In Ireland and the United Kingdom, researchers [75] applied these frameworks to develop an industrial scale strategic energy management system that models, measures and monitors electricity use. Findings from these pilot studies demonstrate improvements in electricity measurement, verification, and control, leading to more efficient use, cost savings and reductions in greenhouse gas emissions [71, 72, 75].

Furthermore, researchers have piloted the use of data from IoT devices, control switches and meteorological information, to increase the performance and power generation of renewable solar electricity generation plants, with a view to improving fault tolerance, resilience, and reliability [76]. Researchers have also piloted the use of IoT devices in biogas generation to centralise the control and monitoring of decentralised, locally situated biogas plants. Resulting in increased efficiency, improved management, and monitoring [77]. Ongoing work is also being undertaken with smart grid technology. These grids use IoT sensors and smart meters to send real-time information signals regarding electricity quality and consumption. Such information allows suppliers to monitor, control and make decisions regarding improved energy flow and increasing electricity grid efficiency [78]. In USA, Germany and Australia, researchers have also been piloting similar technology. Findings indicate that consumers can control and manage their electricity usage through a combination of in-home displays, online monitoring applications and smart meters. Results from these pilots, though tentative, indicate improvements in grid performance and efficiency and a reduction in greenhouse gas emission [79].

7.3.3 IoT's: Societal Health and Well-Being

More recently, much work has been undertaken to identify how IoT devices in combination with digital wearables can improve citizens health and wellbeing, referred to as smart health care. Traditional medical practices, ICT, and IoT devices (e.g., medical equipment, sensors, and wearables) are combined with sensitive patient data, being collected, and shared in real-time across secured networks between healthcare providers and medical specialists, to support real-time decision making by practitioners. This has the potential to improve the efficiency of healthcare provision, reduce costs and improve citizen health and wellbeing [57].

The use of IoT's and digital wearable(s) has also been trialled to monitor the state of health in real-time, allowing early intervention by specialists, for elderly patients [80, 81], and for those with chronic health conditions [82]. Other researchers have identified the benefits from IoT sensors in improving the wellbeing of elderly individuals with cognitive impairment(s) such as dementia [83]. For example, researchers [84] used digital wearable(s) connected to IoT sensors to monitor in real-time the ability of elderly patients, to perform simple and complex tasks while at home. By comparing the data collected against simple and complex activity life scores, medical practitioners were able to intervene earlier, when results indicated that their patients functioning had deteriorated, enabling them to intervene quicker and take action(s) to improve their patient's quality of life.

Commentators have long recognised that traditional hospital environments that adopt donor-centric approaches to chronic disease management, will increasingly become incapable of providing adequate support to an increasing number of patients with chronic disease(s). Given that chronic diseases are often incurable and of long duration, treatment can place increasing burdens on hospital medical personnel and deplete already limited financial resources [85]. Consequentially, researchers have begun exploring how IoT devices can be used to provide improved and informed in-home care for these patients. As such, IoT devices can be situated around patients' homes and connected to digital wearables to act as virtual assistances, to provide medication reminders, send messages and information to and from primary care providers on patient's health, and or to assist disabled patients with simple in-home tasks and activities [82]. Furthermore, researchers have also begun exploring how IoT devices can be used to increase hospital efficiency and improve overall productivity. For example, by using sensors and Radio Frequency Identification (RFID) tags for staff and patient monitoring [86], and tracking inventory safely, reliably, and efficiently [87].

Though informative and rich in data, these case studies are the tip of the iceberg for the work currently being done in exploring the usage of IoT devices in urban settings. Nevertheless, they do demonstrate the enormous potential of IoT device usage and contribution value to a city's environmental sustainability and the societal health and wellbeing of its citizens. However, two questions arise: (1) What are the challenges and issues that limit a more expansive use of IoT's in urban environments by municipal governments? And, (2) What steps can or should be taken to resolve them?

7.4 IoT's: Challenges, Problems, Issues, and Impacts

Several studies have identified a range of technological and social challenges, issues, risks, and obstacles to IoT adoption, often with unintended consequences [88, 89]. For example, adopting IoT devices, as virtual assistants to lower costs and improve efficiency of aged care [82], may inadvertently increase loneliness, isolation and or increasing mental ill-health. Other commentators have become concerned regarding the technological risks associated with the volumes of sensitive personal data and information being exchanged between IoT devices. Concerns being raised regarding the impact(s) on citizens sense of privacy if data leaks reveal their sensitive and personal information [90–93]. It is also identified that societal challenges, raising issues about citizens perceptions of the degree to which they can control how IoT devices use their personal data. For example, White [94], found that when citizens have control over the use of their personal health data, they are more likely to view IoT health care devices more positively. Other challenges and issues include high-implementation costs, lack of appropriate legal frameworks for data and information management and storage, and the complexity inherent in large scale IoT programs [88]. Such challenges have led some researchers to question whether smart cities and their reliance on ICT and IoT's can or do address the issues of rapid urbanisation. For example, the researcher of [95] question(s) the way in which the smart city technology agenda has even been driven, suggesting that through both perceived and imaginary crises, viewed by global technology and consultancy firms, demands for municipal governments to transparently re-evaluate IoT's benefits have ultimately eventuated.

7.4.1 IoT's: Communication and Data Transmission

With minimal human intervention, usage of IoT devices enables the communication and transmission of data and information between cyber and human systems with relative ease. Resulting in potentially billion(s) of inter-connections between smart devices and applications. Commentators have long recognised that several technological challenges and obstacles exist that may inhibit the deployment of IoT on a large scale. IoT devices and their applications require, heterogeneity of connections and objects, unique classifications, and schema to allow them to connect to the internet and communicate and transfer data reliably, uniquely, persistently, and to scale [96]. Currently, used internet protocols have significant limitations in terms of the number of available Internet Protocol (IP) addresses (e.g., IPv4 has reached its limit). To resolve these challenges, researchers are exploring the use of IPv6, which reportedly has an abundance of available addresses [97], is light weight, has virtual domain and multi-encoding [98, 99]. IoT devices also require the collection, exchange, and aggregation of huge volumes of data (i.e., big data), that is often un-structured, and from multiple and disparate sources. This data requires filtering (i.e., removal of unwanted information), and algorithms to enable value extraction and trend(s) analysis to inform decision making [100]. However, wider use of IoT is limited due to a lack of agreement on the definition of big data within the context of IoT, and a lack of consistent evaluation standards and frameworks for quality and algorithm design [101].

7.4.2 IoT's: Data Security and Privacy

Research indicates that user control and confidence, or lack there-of, in information security and privacy greatly impacts the uptake in IoT system usage [102]. User trust ultimately influences their intentions when using new technology, including IoT's [103], and their willingness to share data with third parties [104]. However, the diversity of IoT devices and their communication requirements increase the criticality and complexity associated with security, privacy, and trust concerns [100, 105]. Vulnerabilities in overall system design create opportunities for malware and backdoor software installations for IoT devices and applications. To resolve these challenges, commentators suggest that municipal government(s) need to: enhance their citizens IT security literacy; amend and improve data storage and management regulations; take active steps to re-enforce physical infrastructure [106]; include platforms of security- as-design, corrective software, and hardware patching facilities; and establish security emergency response teams [107]. This requires a shift in focus away from achieving basic functional requirements, to a more holistic approach that upholds and enforces security considerations across and throughout the entire design, deployment, and management of IoT devices [100].

7.4.3 IoT's: Quality of Service (QoS)

Successful end-to-end IoT deployment requires the ability for different service users to send and transmit data with minimal delay. As such, information loss within network bandwidths needs to meet required Quality of Service (QoS) levels. Given the importance of security issues and concerns, research and development remain imperative. To date, while some work has been done to research how best to implement stable and diverse networks of IoT's, further research needs to be undertaken to develop more generalised models targeted at improved and reliable QoS levels [108]. Lastly, a further technological challenge for system designers, is that IoT devices and or their applications, currently rely on battery power and or energy harvesting techniques. As such, when designing an IoT system, consideration needs to be made to minimise the unnecessary transfer of unneeded data and device protocols. Designing such a system requires a rethinking of network architecture and the nature and efficiency of routing mechanisms [109]. With these challenges in mind, researchers have been exploring the role that evolving Fourth and Fifth Generation (4G and 5G) communication platforms may play in the development of high speed, more efficient, more secure, longer battery life, and the reliability of connections between IoT's, smartphones and other devices and applications [110].

7.4.4 IoT's: Social Challenges

Whilst IoT devices operate somewhat semi-autonomously, their applications often require human intervention and interaction to, for example, change an event and or process [100]. Yet challenges emerge with citizen acceptance of IoT device use. Research indicates that poor or lack of communication to citizens regarding the usefulness and benefits of IoT devices remains the overwhelming reason for the slow acceptance rate of IoT related applications [111]. Additionally, urban planners and municipal governments often exclude or neglect citizen involvement, in top-down designs for smart city programs and decision-making pertaining to IoT deployment in their communities. Resultingly, citizens are left feeling discriminated against, resistant to, critical of, and concerned about uninformed use of their personal data and information [112, 113]. Commentators have further argued that IoT's and technology proliferation in smart cities may inadvertently reduce the inference and reasoning capacity of citizens, through unexpected skill loss of the user as they become increasingly dependent and reliant on and lose control over technology use [114].

7.5 Addressing ICT and IoT Problems, Issues and Challenges—A CSR Approach

For evolving smart cities, addressing problems, issues and challenges associated with ICT and IoT usage remain essential. Commentators argue that such a diverse challenge requires policymakers to avoid: adopting a techno-structural view of smart cities and a top-down approach to decision making and ICT and IoT implementation; and, allowing business and private interests to take precedent over the needs of citizens and community groups [2, 43, 115]. Rather, policymakers democratically empower and engage with all stakeholder groups, including business(es) and citizens, at every stage in smart city projects to determine what a smart city is and means, and how it should be shaped. Resultingly, meeting citizen needs further strengthens citizen trust in government [43]. This requires that policymakers work collaboratively with business(es) to identify, understand, and respond to citizens attitudes, behaviours, perspectives and opinions on the problems, challenges, and issues they see arising from ICT and IoT use within smart cities.

As such, governance practices of evolving smart cities require an approach that is simultaneously top-down and bottom-up [2]. In this way, business demonstrates a willingness to not impose ICT and IoT solution(s) on citizens, that address the urban problems that they consider most relevant [2, 43]. As illustrated (see Fig. 7.1), governments working and collaborating with business(es) and citizens in the design and development of smart cities, including the use of IoT's, through a simultaneous top-down/bottom-up approach requires engaging with and communicating with multiple stakeholders across three layers of activity. The presented model poses the argument that each interactive layer is porous, in that feedback and feedforward loops in and between each layer generate the identification of problems, issues and challenges, and potential solutions that feedback into the next and vice versa. The outer layer captures the city macro-environmental factors, at a financial, social, human, and environmental level(s) that have impacts and implications for and on the dynamic and evolving nature of city(ies).

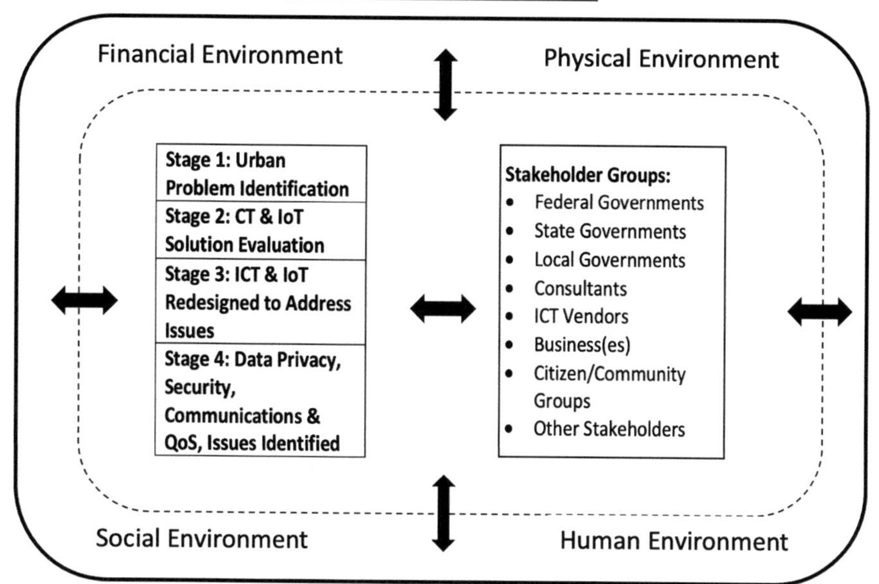

Fig. 7.1 Smart city design: simultaneous top-down/bottom-up Governance model

The middle layer is where smart city solutions that are seen as relevant to the resolution of macro-environmental problems are identified, considered, and deliberated by collective stakeholder groups. Once agreed upon by stakeholders, specific urban problems are evaluated against potential ICT and IoT solutions. Here the specific solutions undergo detailed evaluation through evolving, cyclical, iterative and dynamic process(es) of interactions by multiple stakeholder groups. During this process, consideration is given to the impacts and implications of the emergent ICT and IoT data privacy and security issues, QoS, and network and communication(s) challenges. Much collaborative work is undertaken to resolve these challenges, including identifying and implementing the necessary steps to communicate how these issues and problems are and will be addressed, before and throughout the deployment, implementation and delivery of intended technology(ies).

This simultaneous top-down/bottom-up approach of engagement, communication and collaboration by multiple stakeholder groups/governance model, begins at the inner layer (see Fig. 7.1), involving four stages that also operate asynchronously and somewhat interactively:

Stage (1) Urban Problem Identification: Here businesses and government collaborate with stakeholders involving them in discussion(s) and debate(s) on their views on urban problems. Policy, procedural and governmental challenges are transparently shared. Discussions also take place about the role of ICT and IoT in increasing efficiencies and productivity in city infrastructure provision.

Stage (2) ICT & IoT solution evaluation: Here asynchronous collaboration occurs between ICT vendors, consultants, governments, business(es) and citizens about the various ICT and IoT solutions available for urban problem resolution. These solutions are compared with an emphasis placed upon the degree to which they do or do not resolve the urban problems identified in *Stage 1*. Stakeholders work together to identify and debate the negative and positive externalities perceived by each solution and develop a pathway to negative externalities minimisation. Returning to *Stage 1*, if a more detailed understanding and or reclarification of the urban problem is required.

Stage (3) Privacy, Security, Communications & QoS Issue Identification: In this stage, businesses, including ICT vendors and consultants, and citizens collaborate to identify, evaluate, and seek to resolve the privacy and security concerns raised by the public. Communication and QoS problems are also debated and addressed.

Stage (4) ICT Solution Redesign to Address Issues: Here the impacts and implications of the issues identified in *Stage 3* are compared against current and future technology approaches. With business(es) identifying way(s) in ICT and IoT solutions could be re-designed to both address the issues and the steps they can take to address community and citizens' concerns, in ways that demonstrate social responsibility and responsiveness. Businesses also evaluate their CSR strategies taking steps to build upon and extend them to alter their approaches to technology and IoT solution use, including revised performance measures, in ways that are contextually relevant to the citizens' group(s) most impacted [2, 47].

This approach to the use of ICT and IoT in smart city projects shifts the emphasis away from autocratic business and governmental decision making and its emphasis on techno-structural approaches [43] and prevents business goal(s) from being prioritised over the social, economic, cultural, and environmental needs of other stakeholder groups, including citizens [43, 116]. Further, the ideas and suggestions arising from each stage(s) of collaboration enable the CSV by policymakers, businesses, and all stakeholder groups. Policymakers and business(es) can use the knowledge and information they harness from each stage to both develop a greater insight and understanding of how citizens use and engage with ICT and IoT devices. Such understandings and insights can be further used to enable open innovation, enhance entrepreneurial activities and facilitate innovation to create new and more sustainable business models [2, 117, 118].

7.6 Discussion

While the need for ongoing and future research regarding smart city(ies) development and the impact of, or reliance on, benefits and challenges of ICT and IoT's remains uncontested, this chapter also recognises the plethora of research undertakings and interest over the last few decades, whereby scholars have and will continue to be increasingly interested in exploring the use of ICT and IoT devices within smart city(ies) environments [48, 49]. Such interest, triggered by the potential of smart technology(ies), to address the economic, environmental and social problems associated with increased urbanisation and rapid population growth [2, 11–19], and a drive towards making our cities more environmentally sustainable [50, 57]. To date, significant work has been done to demonstrate how IoT devices can improve; waste [51, 52], water [53], air pollution [54], urban transport [56], electricity supply [9, 55], and management, while also making significant contributions to societal health and wellbeing [10]. However, scholars have also criticised the over-reliance on smart technologies (e.g., networked IoT devices) by urban planners [36] and raised concerns about the dominance of global corporations (e.g., IBM, CISCO, and KPMG) in the smart city technology(ies) transformation process. Scholars question the way policymakers over emphasise business's needs and give them a dominant position, including undue power and influence within smart city agenda(s) settings, including areas of decision making, problem(s) identification, and resolution [2, 36, 43]. Such positions, given to the detriment of the needs of citizens, especially those from diverse, ethnic, and religious backgrounds [25, 26]. This chapter argues such challenges become compounded when policymakers add IoT devices to the smart technology mix.

IoT device(s) generate their own challenges, problems, and issues that policymakers need to address both now and well into the future of smart city development [3–11]. Concerns have been raised about data and information privacy [90–93], and the associated caution and mistrust by citizens over IoT device(s) use within communities [100, 105], and amongst others. These concerns create inevitable obstacles to the potential uptake of IoT usage by citizens [102]. This chapter explores these challenges and issues in some detail.

Adopting a CSR and CSV perspective(s), and viewing city(ies) as evolving, interconnected systems of systems [2, 33], this chapter builds a case for the adoption of a multi-stakeholder CSR approach to the IoT design and decision-making processes within the smart city(ies). An approach that avoids singularly focused top-down approaches to decision making and solution implementation by policymakers. Rather, addressing the issues and challenges associated with ICT and IoT adoption within smart cities, through recognising and arguing the wisdom of a top-down and bottom-up management approach, and the importance of CSV. Doing so relies on policymakers working collaboratively with, engaging and empowering all city stakeholder(s), including businesses, community groups and citizens in the determination of urban problems, their solution(s), and the selection and deployment of IoT devices [2, 43, 115].

7.7 Conclusion

This chapter argues that working collaboratively to identify, understand, and respond to citizens attitudes and perspectives of IoT's, increases citizen trust in government and strengthens their understanding, awareness, and acceptance of IoT devices. The chapter concludes by proposing a conceptual model that places citizens and their communities at the centre of an iterative, evolving multi-stage collaborative process. This shift requires actively involving all citizen groups, including those from diverse, ethnic, and religious backgrounds, in debating the challenges and problems faced in their communities and discussing and addressing the risks and issues they see emerging from IoT devices and ICT. Whilst not necessarily enabling the resolution of all the IoT device issues, given that some are driven by current technology limitations, this chapter argues the viewpoint that such a collaborative approach encourages(s) future and ongoing multi-stakeholder dialogue and open innovation to identify and address, and or minimise the risks identified regarding IoT device(s) and their use,

References

1. Tanda, A., De Marco, A.: Drivers of public demand of IoT-enabled smart city services: a regional analysis. J. Urban Technol. **25**(4), 77–94 (2018). https://doi.org/10.1080/10630732.2018.1509585
2. Roberts, A.D.: Corporate social responsibility (CSR): governments, institutions, businesses, and the public within a smart city context. In: Handbook of Smart Cities, pp. 1–25 (2020). https://doi.org/10.1007/978-3-030-15145-4_30-1
3. Zhang, X.Q.: The trends, promises and challenges of urbanisation in the world. Habitat Int **54**(Part 3), 241252 (2016). ISSN: 0197-3975, 1873-5428. https://doi.org/10.1016/j.habitatint.2015.11.01
4. Bai, X., Nath, I., Capon, A., Hasan, N., Jaron, D.: Health and wellbeing in the changing urban environment: complex challenges, scientific responses, and the way forward. Current opinion in environmental sustainability. **4**(4), 465–472 (2012). https://doi.org/10.1016/j.cosust.2012.09.009
5. World Bank: What a Waste' Report Shows Alarming Rise in Amount, Cost of Garbage. World Bank (2012)
6. González-Vidal, A., Cuenca-Jara, J., Skarmeta, A.F.: IoT for water management: towards intelligent anomaly detection. In: 2019 IEEE 5th World Forum on Internet of Things (WF-IoT), pp. 858–863, Apr 15 (2019). IEEE. ISBN: 9781538649800
7. Kennedy, C., Cuddihy, J., Engel-Yan, J.: The changing metabolism of cities. J. Ind. Ecol. **11**(2), 43–59 (2007). https://doi.org/10.1162/jie.2007.1107
8. Wright, L., Fulton, L.: Climate change mitigation and transport in developing nations. Transp. Rev. **25**(6), 691–717 (2005). https://doi.org/10.1080/01441640500360951
9. da Silva, F.S.T., da Costa, C.A., Crovato, C.D.P., da Rosa Righi, R.: Looking at energy through the lens of industry 4.0: a systematic literature review of concerns and challenges. Comput. Ind. Eng. **143**, 106426 (2020). https://doi.org/10.1016/j.cie.2020.106426
10. Vallée, T., Sedki, K., Despres, S., Jaulant, M.C., Tabia, K., Ugon, A.: On personalization in IoT. In: 2016 International Conference on Computational Science and Computational Intelligence (CSCI), pp. 186–191, Dec 15 (2016). IEEE. https://doi.org/10.1109/CSCI.2016.0042
11. Kummitha, R.K., Crutzen, N.: How do we understand smart cities? An evolutionary perspective. Cities **67**, 43–52 (2017). ISSN: 0264-2751, 1873-6084. https://doi.org/10.1016/j.cities.2017.04.010
12. Feroz, A.K., Zo, H., Chiravuri, A.: Digital transformation and environmental sustainability: a review and research agenda. Sustainability **13**(3), 1530 (2021). https://doi.org/10.3390/su13031530
13. Bokolo, A. Jr.: Managing digital transformation of smart cities through enterprise architecture–a review and research agenda. Enterp. Inf. Syst. **15**(3), 299–331 (2021). https://doi.org/10.1080/17517575.2020.1812006

14. Bosdriesz, Y., Van Sinderen, M., Iacob, M., Verkroost P.: A building information model-centered big data platform to support transformation in the construction industry. In: Enterprise Interoperability: Smart Services and Business Impact of Enterprise Interoperability, pp. 209–215. Wiley, London, UK (2018). https://doi.org/10.1002/9781119564034.ch26
15. Salem, F.A.: Smart city for public value: digital transformation through agile governance-the case of 'Smart Dubai'. World Government Summit Publications (2016)
16. Bibri, S.E.: The IoT for smart sustainable cities of the future: an analytical framework for sensor-based big data applications for environmental sustainability. Sustain. Cities Soc. **38**, 230–253 (2018). https://doi.org/10.1016/j.scs.2017.12.034
17. Malik, K.R., Sam, Y., Hussain, M., Abuarqoub, A.A.: Methodology for real-time data sustainability in smart city: towards inferencing and analytics for big-data. Sustain. Cities Soc. **39**, 548–556 (2018). https://doi.org/10.1016/j.scs.2017.11.031
18. Landry, C.: The art of city making. Routledge, London (2006). https://doi.org/10.4324/9781849772877
19. Zhang, Z.K., Cho, M.C., Wang, C.W., Hsu, C.W., Chen, C.K., Shieh, S.: IoT security: ongoing challenges and research opportunities. In: IEEE 7th International Conference on Service-Oriented Computing and Applications, pp. 230–234. IEEE (2014). https://doi.org/10.1109/SOCA.2014.58
20. Drepaul, N.A.: Sustainable cities and the Internet of Things (IOT) technology. Consilience **6**(22), 39–47 (2020). https://doi.org/10.2307/26924960
21. Komninos, N., Schaffers, H., Pallot, M.: Developing a policy roadmap for smart cities and the future internet. In eChallenges e-2011 Conference Proceedings, IIMC International Information Management Corporation., IMC International Information Management Corporation (2011). ISBN: 9781905824274
22. Kortuem, G., Kawsar, F., Sundramoorthy, V., Fitton, D.: Smart objects as building blocks for the internet of things. IEEE Internet Comput. **14**(1), 44–51 (2009). https://doi.org/10.1109/MIC.2009.143
23. Hinze, A., Sachs, K., Buchmann, A.: Event-based applications and enabling technologies. In: Proceedings of the Third ACM International Conference on Distributed Event-Based Systems, pp. 1–15 (2009). https://doi.org/10.1145/1619258.1619260
24. Leach, J., Mulhall, R., Rogers, C., Bryson, J.: Reading cities: developing an urban diagnostics approach for identifying integrated urban problems with application to the city of Birmingham, UK. Cities **86**, 136–144 (2019). ISSN: 0264-2751, 1873-6084. https://doi.org/10.1016/j.cities.2018.09.012
25. Meijer, A., Bolívar, M.P.: Governing the smart city: a review of the literature on smart urban governance. Int. Rev. Adm. Sci. **82**(2), 392–408 (2016). https://doi.org/10.1177/0020852314564308
26. Boyle, M.E., Boguslaw, J.: Business, poverty and corporate citizenship: naming the issues and framing solutions. J. Corp. Citizsh. **26**, 101–120 (2007)
27. Landry C.: The Creative City: A Toolkit for Urban Innovators. Routledge, New York (2012). ISBN: 978-1-84407-598-0
28. Wood, P., Landry, C.: The Intercultural City: Planning for Diversity Advantage. Routledge, New York (2008). ISBN: 978-1-84407-437-2
29. Letaifa, S.: How to strategize smart cities: revealing the SMART model. J. Bus. Res. **68**(7), 1414–1419 (2015). ISSN: 0148-2963, 1873-7978. https://doi.org/10.1016/j.jbusres.2015.01.024
30. Cugurullo, F.: How to build a sandcastle: an analysis of the genesis and development of Masdar city. J. Urban Technol. **20**(1), 23–37 (2013). https://doi.org/10.1080/10630732.2012.735105
31. Healey, P.: Urban Complexity and Spatial Strategies: Towards a Relational Planning for Our Times. Routledge, Abingdon (2007). https://doi.org/10.4324/9780203099414
32. Cosgrave, E., Arbuthnot, K., Tryfonas, T.: Living labs, innovation districts and information marketplaces: a systems approach for smart cities. Procedia Comput. Sci. **16**, 668–677 (2013). https://doi.org/10.1016/j.procs.2013.01.070

33. Merrilees, B., Miller, D., Herington, C.: Antecedents of residents' city brand attitudes. J. Bus. Res. **62**(3), 362–367 (2009). ISSN: 0148-2963, 1873-7978. https://doi.org/10.1016/j.jbusres.2008.05.011
34. Söderström, O., Paasche, T., Klauser, F.: Smart cities as corporate storytelling. City **18**(3), 307–320 (2014). https://doi.org/10.1080/13604813.2014.906716
35. Cowley, R., Joss, S., Dayot, Y.: The smart city and its publics: insights from across six UK cities. Urban Res. Pract. **11**(1), 53–77 (2018). https://doi.org/10.1080/17535069
36. Vanolo, A.: Is there anybody out there? The place and role of citizens in tomorrow's smart cities. Futures **82**, 26–36 (2016). ISSN: 0016-3287, 1873-6378. https://doi.org/10.1016/j.futures.2016.05.010
37. Kitchin, R.: The real-time city? Big data and smart urbanism. GeoJournal **79**, 1–14 (2014). https://doi.org/10.1007/s10708-013-9516-8
38. Yigitcanlar, T.: Technology and the City: Systems, Applications and Implications. Routledge, New York (2016). https://doi.org/10.4324/9781315739090
39. Powell, A.: Datafication, transparency, and good governance of the Data City. In: O'Hara, M.H.K., Nguyen, C., Haynes, P. (eds.) Digital Enlightenment Yearbook, pp. 215–224. IOS Press BV, Amsterdam (2014)
40. Saunders, T., Baeck, P.: Rethinking smart cities from the ground up. Nesta, London (2015). www.nesta.org.uk
41. Wood, D.J.: Measuring corporate social performance: a review. Int. J. Manage. Rev. **12**(1), 50–84 (2010). https://doi.org/10.1111/j.1468-2370.2009.00274.x
42. Hollands, R.: Will the real smart city please stand up? Creative, progressive or just entrepreneurial? City: Anal. Urban Trends Culture Theory Policy Action **12**, 303–320 (2008). ISSN: 1360-4813, 1470-3629. https://doi.org/10.1080/13604810802479126
43. Carroll, A.B.: A three-dimensional conceptual model of corporate social performance. Acad. Manage. Rev. **4**, 497–505 (1979). ISSN: 0363-7425, 1930-3807. https://doi.org/10.5465/amr.1979.4498296
44. Carroll, A.B.: A history of corporate social responsibility: concepts and practices. In: Crane, A., McWilliams, A., Matten, D., Moon, J., Siegel, D. (eds.) The Oxford Handbook of Corporate Social Responsibility, vol. 1, pp. 19–46 (2008). https://doi.org/10.1093/oxfordhb/9780199211593.003.0002
45. Porter, M.E., Kramer, M.R.: Creating shared value. Harv. Bus. Rev. **89**, 62–77 (2011). https://doi.org/10.1007/978-94-024-1144-7_16
46. Crane, A., Palazzo, G., Spence, L.J., Matten, D.: Contesting the value of the shared value concept. Calif. Manage. Rev. **56**(2), 130–153 (2014). https://doi.org/10.1525/cmr.2014.56.2.130
47. Botta, A., De Donato, W., Persico, V., Pescapé, A.: Integration of cloud computing and internet of things: a survey. Future Gener. Comput. Syst. **56**, 684–700 (2016). https://doi.org/10.1016/j.future.2015.09.021
48. Rose, K., Eldridge, S., Chapin, L.: The internet of things: an overview. The Internet Society (ISOC) **80**, 1–50 (2015). www.internetsociety.org
49. Pawar, A., Kolte, A., Sangvikar, B.: Techno-managerial implications towards communication in internet of things for smart cities. Int. J. Pervasive Comput. Commun. (2021). https://doi.org/10.1108/IJPCC-08-2020-0117
50. Ismail, N.A., Ab Majid, N.A., Hassan, S.A.: IoT-based smart solid waste management system: a systematic literature review. Int. J. Innov. Technol. Explor. Eng. (IJITEE) 2278–3075 (2019). ISSN: 2278-3075
51. Pardini, K., Rodrigues, J.J., Kozlov, S.A., Kumar, N., Furtado, V.: IoT-based solid waste management solutions: a survey. J. Sens. Actuator Netw. **8**(1), 5 (2019). https://doi.org/10.3390/jsan8010005
52. Sharma, A., Sharma, R.: A review of applications, approaches, and challenges in internet of things (IoT). Proceedings of ICRIC, pp. 257–269 (2020). https://doi.org/10.1007/978-3-030-29407-6_20

53. Idrees, Z., Zou, Z., Zheng, L.: Edge computing based IoT architecture for low cost air pollution monitoring systems: a comprehensive system analysis, design considerations & development. Sensors **18**(9), 3021 (2018). https://doi.org/10.3390/s18093021
54. Lahti, J.P., Helo, P., Shamsuzzoha, A., Phusavat, K.: IoT in electricity supply chain: review and evaluation. In: 2017 15th International Conference on ICT and Knowledge Engineering (ICT&KE), pp. 1–6. IEEE (2017). https://doi.org/10.1109/ICTKE.2017.8259615
55. Araújo, A., Kalebe, R., Giraõ, G., Gonçalves, K., Neto, B.: Reliability analysis of an IoT-based smart parking application for smart cities. In: 2017 International Conference on Big Data (Big Data), pp. 4086–4091. IEEE (2017). https://doi.org/10.1109/BigData.2017.8258426
56. da Silva, B.N., Khan, M., Han, K.: Futuristic sustainable energy management in smart environments: a review of peak load shaving and demand response strategies, challenges, and opportunities. Sustainability. **12**(14), 5561 (2020). https://doi.org/10.1016/j.cie.2020.106426
57. Thakker, S., Narayanamoorthi, R.: Smart and wireless waste management. In: International Conference on Innovations in Information, Embedded and Communication Systems (ICIIECS), pp. 1–4. IEEE (2015). https://doi.org/10.1109/ICIIECS.2015.7193141
58. Catania, V., Ventura, D.: An approach for monitoring and smart planning of urban solid waste management using smart-M3 platform. In: Proceedings of the 15th Conference of Open Innovations Association FRUCT, pp. 24–31. IEEE, April 21 (2014). https://doi.org/10.1109/FRUCT.2014.6872422
59. Ismail, M.I., Dziyauddin, R.A., Salleh, N.A., Muhammad-Sukki, F., Bani, N.A., Izhar, M.A., Latiff, L.A.: A review of vibration detection methods using accelerometer sensors for water pipeline leakage. IEEE Access **7**, 51965–51981 (2019). ISSN: 2278-3075
60. Giaquinto, N., Cataldo, A., D'Aucelli, G.M., De Benedetto, E., Cannazza, G.: Water detection using bi-wires as sensing elements: comparison between capacimetry-based and time-of-flight-based techniques. IEEE Sens. J. **16**(114), 309–317 (2016). https://doi.org/10.1109/JSEN.2016.2540299
61. Koo, D., Piratla, K., Matthews, C.J.: Towards sustainable water supply: schematic development of big data collection using internet of things (IoT). Procedia Eng. **118**, 489–497 (2015). https://doi.org/10.1016/j.proeng.2015.08.465
62. Telstra (2021). Available at: https://www.telstra.com.au/aboutus/media/media-releases/telstra-yarra-valley-water-iot
63. Alshamsi, A., Anwar, Y., Almulla, M., Aldohoori, M., Hamad, N., Awad, M.: Monitoring pollution: applying IoT to create a smart environment. In: 2017 International Conference on Electrical and Computing Technologies and Applications (ICECTA), pp. 1–4. IEEE (2017). https://doi.org/10.1109/ICECTA.2017.8251998
64. Kanabkaew, T., Mekbungwan, P., Raksakietisak, S., Kanchanasut, K.: Detection of PM2.5 plume movement from IoT ground level monitoring data. Environ. Pollut. **252**, 543–552 (2019). https://doi.org/10.1016/j.envpol.2019.05.082
65. Suciu, G., Balanescu, M., Nadrag, C., Birdici, A., Balaceanu, C.M., Dobrea, M.A., Pasat, A., Ciobanu, R.I.: IoT system for air pollutants assessment in underground infrastructures. In: Proceedings of the 6th Conference on the Engineering of Computer Based Systems, pp. 1–7, Sep 2 (2019). https://doi.org/10.1145/3352700.3352710
66. Arco, E., Boccardo, P., Gandino, F., Lingua, A., Noardo, F., Rebaudengo, M.: An integrated approach for pollution monitoring: smart acquirement and smart information. ISPRS Ann. Photogrammetry Remote Sens. Spatial Inf. Sci. **3**(1) (2016). https://doi.org/10.5194/isprs-annals-IV-4-W1-67-2016
67. Nahar, S.A.A., Hashim, F.H.: Modelling and analysis of an efficient traffic network using ant colony optimization algorithm. In: 2011 Third International Conference on Computational Intelligence, Communication Systems and Networks (CICSyN), pp. 32–36. IEEE (2011). https://doi.org/10.1109/CICSyN.2011.20
68. Kponyo, J.J., Kuang, Y., Li, Z.: Real time status collection and dynamic vehicular traffic control using ant colony optimization. In: 2012 International Conference on Computational Problem-Solving (ICCP), pp. 69–72. IEEE (2012). https://doi.org/10.1109/ICCPS.2012.6384297

69. Sang, K.S., Zhou, B., Yang, P., Yang, Z.: Study of group route optimization for IoT enabled urban transportation network. In: 2017 IEEE International Conference on Internet of Things (iThings) and IEEE Green Computing and Communications (GreenCom) and IEEE Cyber, Physical and Social Computing (CPSCom) and IEEE Smart Data (SmartData), pp. 888–893, Jun 21 (2017). https://doi.org/10.1109/iThings-GreenCom-CPSCom-SmartData.2017.137
70. Piyare, R., Tazil, M.: Bluetooth based home automation system using cell phone. In: 2011 IEEE 15th International Symposium on Consumer Electronics (ISCE), pp. 192–195. IEEE (2011). https://doi.org/10.1109/ISCE.2011.5973811
71. Patil, K., Metan, J., Kumaran, T.S., Mathapatil, M.: IoT based power management and controlled socket. In: 2017 International Conference on Electrical, Electronics, Communication, Computer, and Optimization Techniques (ICEECCOT), pp. 243–247. IEEE (2017). ISBN: 9781538623619
72. Metallidou, C.K., Psannis, K.E., Egyptiadou, E.A.: Energy efficiency in smart buildings: IoT approaches. IEEE Access **8**, 63679–63699 (2020). https://doi.org/10.1109/ACCESS.2020.2984461
73. Wei, M., Hong, S.H., Alam, M.: An IoT-based energy-management platform for industrial facilities. Appl. Energy **164**, 607–619 (2016). https://doi.org/10.1016/j.apenergy.2015.11.107
74. Swords, B., Coyle, E., Norton, B.: An enterprise energy-information system. Appl Energy **85**(1), 61–69 (2008). https://doi.org/10.1016/j.apenergy.2007.06.009
75. Phung, M.D., De La Villefromoy, M., Ha, Q.: Management of solar energy in microgrids using IoT-based dependable control. In: 2017 20th International Conference on Electrical Machines and Systems (ICEMS), pp. 1–6. IEEE (2017). ISBN: 9781538632468
76. Logan, M., Safi, M., Lens, P., Visvanathan, C.: Investigating the performance of internet of things based anaerobic digestion of food waste. Process Saf. Environ. Prot. **127**, 277–287 (2019). https://doi.org/10.1016/j.psep.2019.05.025
77. Morello, R., De Capua, C., Fulco, G., Mukhopadhyay, S.C.: A smart power meter to monitor energy flow in smart grids: the role of advanced sensing and IoT in the electric grid of the future. IEEE Sens. J. **17**(23), 7828–7837 (2017). https://doi.org/10.1109/JSEN.2017.2760014
78. Haidar, A.M., Muttaqi, K., Sutanto, D.: Smart Grid and its future perspectives in Australia. Renew. Sustain. Energy Rev. **51**, 1375–1389 (2015). https://doi.org/10.1016/j.rser.2015.07.040
79. Ahad, A., Tahir, M., Aman, S.M., Ahmed, K.I., Mughees, A., Numani, A.: Technologies trend towards 5G network for smart health-care using IoT: a review. Sensors **20**(14), 4047 (2020). https://doi.org/10.3390/s20144047
80. Ullah, K., Shah, M.A., Zhang, S.: Effective ways to use Internet of Things in the field of medical and smart health care. In: 2016 International conference on intelligent systems engineering (ICISE), pp. 372–379. IEEE (2016). https://doi.org/10.1109/INTELSE.2016.7475151
81. Tian, S., Yang, W., Le Grange, J.M., Wang, P., Huang, W., Ye, Z.: Smart healthcare: Making medical care more intelligent. Global Health J. **3**(3), 62–65 (2019). https://doi.org/10.1016/j.glohj.2019.07.001
82. De-La-Hoz-Franco, E., Ariza-Colpas, P., Quero, J.M., Espinilla, M.: Sensor-based datasets for human activity recognition–a systematic review of literature. IEEE Access **6**, 59192–59210 (2018). https://doi.org/10.1109/ACCESS.2018.2873502
83. Javed, A.R., Fahad, L.G., Farhan, A.A., Abbas, S., Srivastava, G., Parizi, R.M., Khan, M.S.: Automated cognitive health assessment in smart homes using machine learning. Sustain. Cities Soc. **2** (2021). https://doi.org/10.1016/j.scs.2020.102572
84. Willard-Grace, R., DeVore, D., Chen, E.H., Hessler, D., Bodenheimer, T., Thom, D.H.: The effectiveness of medical assistant health coaching for low-income patients with uncontrolled diabetes, hypertension, and hyper-lipidemia: protocol for a randomized controlled trial and baseline characteristics of the study population. BMC Family Practice **14**, 27 (2013). http://www.biomedcentral.com/1471-2296/14/27
85. Fuhrer, P., Guinard, D.: Building a smart hospital using RFID technologies. In: European Conference on eHealth. Gesellschaft für Informatik eV (2006). ISBN: 9783885791850

86. Moslehpour, S., Jenab, K., Namburi, N.: Smart RFID based design for inventory management in health care. Int. J. Ind. Eng. Prod. Res. **22**(4), 231–236 (2011). http://ijiepr.iust.ac.ir/article-1-365-en.html
87. Brous, P., Janssen, M., Herder, P.: The dual effects of the Internet of Things (IoT): a systematic review of the benefits and risks of IoT adoption by organizations. Int. J. Inf. Manage. **51**, 101952 (2020). https://doi.org/10.1016/j.ijinfomgt.2019.05.008
88. Shayan, S., Kim, K.P., Ma, T., Nguyen, T.H.: The first two decades of smart city research from a risk perspective. Sustainability. **12**(21), 9280 (2020). https://doi.org/10.3390/su1221 9280
89. Fan, P.F., Wang, L.L., Zhang, S.Y., Lin, T.T.: The research on the internet of things industry chain for barriers and solutions. Appl. Mech. Mater. **441**, 1030–1035 (2014). https://doi.org/10.4028/www.scientific.net/AMM.441.1030
90. Hossain, M.A., Dwivedi, Y.K.: What improves citizens' privacy perceptions toward RFID technology? A cross-country investigation using mixed method approach. Int. J. Inf. Manage. **34**(6), 711–719 (2014). https://doi.org/10.1016/j.ijinfomgt.2014.07.002
91. Hummen, R., Henze, M., Catrein, D., Wehrle, K.: A cloud design for user-controlled storage and processing of sensor data. In: 4th IEEE International Conference on Cloud Computing Technology and Science Proceedings., pp. 232–240. IEEE (2012). https://doi.org/10.1109/CloudCom.2012.6427523
92. Skarmeta, A.F., Hernandez-Ramos, J.L., Moreno, M.V.: A decentralized approach for security and privacy challenges in the internet of things. In: 2014 IEEE World Forum on Internet of Things (WF-IoT), pp. 67–72. IEEE (2014). https://doi.org/10.1109/WF-IoT.2014.6803122
93. Princi, E., Krämer, N.: I Spy with my little sensor eye-effect of data-tracking and convenience on the intention to use smart technology. In: Proceedings of the 53rd Hawaii International Conference on System Sciences, Jan 7 (2020). https://doi.org/10.24251/HICSS.2020.171
94. White, J.M.: Anticipatory logics of the smart city's global imaginary. Urban Geogr. **37**(4), 572–589 (2016). https://doi.org/10.1080/02723638
95. Gubbi, J., Buyya, R., Marusic, S., Palaniswami, M.: Internet of Things (IoT): a vision, architectural elements, and future directions. Future Gener. Comput. Syst. **29**(7), 1645–1660 (2013). https://doi.org/10.1016/j.future.2013.01.010
96. Palattella, M.R., Accettura, N., Vilajosana, X., Watteyne, T., Grieco, L.A., Boggia, G., Dohler, M.: Standardized protocol stack for the internet of (important) things. IEEE Commun. Surv. Tutorials **15**(3), 1389–1406 (2012). https://doi.org/10.1109/SURV.2012.111412.00158
97. Luo, B., Sun, Z.: Research on the model of a lightweight resource addressing. Chin. J. Electron. **24**(4), 832–836 (2015). https://doi.org/10.1049/cje.2015.10.028
98. Ma, R., Liu, Y., Shan, C., Zhao, X.L., Wang, X.A.: Research on identification and addressing of the internet of things. In: 2015 10th International Conference on P2P, Parallel, Grid, Cloud and Internet Computing (3PGCIC), pp. 810–814. IEEE (2015). https://doi.org/10.1109/3PG CIC.2015.40
99. Farhan, L., Kharel, R., Kaiwartya, O., Quiroz-Castellanos, M., Alissa, A., Abdulsalam, M.: A concise review on Internet of Things (IoT)-problems, challenges and opportunities. In: 2018 11th International Symposium on Communication Systems, Networks & Digital Signal Processing (CSNDSP), pp. 1–6. IEEE (2018). https://doi.org/10.1109/CSNDSP.2018.847 1762
100. Chen, M., Mao, S., Zhang, Y., Leung, V.C.: Big Data: Related Technologies, Challenges and Future Prospects (2014). ISBN: 978-3-319-06245-7
101. Fang, X.W., Chan, S., Brzezinski, J., Xu, S.: Moderating effects of task type on wireless technology acceptance. J. Inf. Manage. Syst. **22**(3), 123–157 (2005). https://doi.org/10.2753/MIS0742-1222220305
102. Kowastasch, T., Maas, W.: Critical privacy factors of internet of things services: an empirical investigation with domain experts. In: Knowledge and Technologies in Innovative information Systems, pp. 200–211. Springer, Berlin Heidelberg (2012). https://doi.org/10.1007/978-3-642-33244-9_14

103. Beales, Muris: Privacy and consumer control. George Mason Law & Economics Research Paper, pp. 19–27 (2019). https://doi.org/10.2139/ssrn.3449242
104. Zhang, Z.K., Cho, M.C., Wang, C.W., Hsu, C.W., Chen, C.K., Shieh, S.: IoT security: ongoing challenges and research opportunities. In: 2014 IEEE 7th International Conference on Service-Oriented Computing and Applications, pp. 230–234. IEEE, Nov 17 (2014). https://doi.org/10.1109/SOCA.2014.58
105. Zhu, Y., Zou, J.: Research on security construction of smart city. Int. J. Smart Home **9**, 197–204 (2015). https://doi.org/10.14257/ijsh.2015.9.8.21
106. Kitchin, R., Dodge, M.: The (in)security of smart cities: vulnerabilities, risks, mitigation and prevention during enterprise systems integration. J. Urban Technol. **26**, 47–65 (2019). https://doi.org/10.1007/s10708-013-9516-8
107. Brogi, A., Forti, S.: QoS-aware deployment of IoT systems through the Fog. IEEE IoT J. **4**(5), 1185–1192 (2017). https://doi.org/10.1109/JIOT.2017.2701408
108. Farahan, L., Kharel, R., Kaiwartya, O., Hammoudeh, M., Abebisi, B.: Towards Green computing for internet of things: energy oriented path and message scheduling approach. Sustain. Cities Soc. **38**, 195–204 (2018). https://doi.org/10.1016/j.scs.2017.12.018
109. Yassein, M.B., Aljawarneh, S., Al-Sadi, A.: Challenges and features of IoT communications in 5G networks. In: 2017 International Conference on Electrical and Computing Technologies and Application (ICECTA). IEEE, Nov 21 (2017). https://doi.org/10.1109/ICECTA.2017.8251989
110. Guo, L., Bai, X.: A unified perspective on the factors influencing consumer acceptance of internet of things technology. Asia Pac. J. Market. Logistics 26(2), 211–231 (2014). ISSN: 1355-5855
111. Gladon-Clavell, G.: (Not so) smart cities? The drivers, impact and risks of surveillance-enabled smart environments. Sci Public Policy **40**, 717–723 (2013). https://doi.org/10.1093/scipol/sct070
112. Trivellato, B.: How can 'smart' also be socially sustainable? In: Insights from the Case of Milan European Urban Regeneration Studies, vol. 24, pp. 337–351 (2017). https://doi.org/10.1177/0969776416661016
113. McGuire, M.: Beyond flatland: when smart cities make stupid citizens. City Territ. Archit. **5**, 22 (2018). https://doi.org/10.1186/s40410-018-0098-0
114. Harvey, D.: From managerialism to entrepreneurialism: the transformation in urban governance in late capitalism. Geografiska Annaler: Series B Human Geogr. **71**(1), 3–17 (1989). ISSN: 0435-3684, 1468-0467. https://doi.org/10.1080/04353684.1989.11879583
115. Brenner, N., Theodeore, N.: Cities and the geographies of "actually existing neoliberalism". Antipode **34**(3), 349–379 (2002). ISSN: 0066-4812, 1467-8330. https://doi.org/10.1111/1467-8330.00246
116. Paskaleva, K.A.: The smart city: a nexus for open innovation? Intell. Build. Int. **3**(3), 153–171 (2011). https://doi.org/10.1080/17508975.2011.586672
117. Visser, W., Kymal, C.: Integrated value creation (IVC): beyond corporate social responsibility (CSR) and creating shared value (CSV). J. Int. Bus. Ethics **8**(1), 29 (2015). ISSN: 1940-1485
118. Coe, A., Paquet, G., Roy, J.: E-governance and smart communities: a social learning challenge. Soc. Sci. Comput. Rev. **19**(1), 80–93 (2001). https://doi.org/10.1177/089443930101900107

Chapter 8
Intelligent Techniques for Optimization, Modelling and Control of Power Management Systems Efficiency

Luis Alfonso Fernandez-Serantes, José-Luis Casteleiro-Roca, and José Luis Calvo-Rolle

Abstract This research analyzes the issue of the climate change and smart grid solutions from the point of view of the efficiency in the field of the power electronics. The power distribution of the Spanish electric system has changed over the years. Moreover, it will further change with the introduction of new concepts, like the smart grids, opening new possibilities for the distribution of energy. In this case, the power electronics aims to interconnect the different parts of the electric system and controlling the energy that flows from point to point. Thus, the optimization, modelling and control of the power electronics converters have become a trending topic nowadays. With the introduction of new technologies almost in every field, the intelligent techniques can be widely used in optimization and control of the power converters, improving the efficiency and control. A classification algorithm to detect the operation mode based on intelligent techniques has been developed and implemented. This intelligent model is able to differentiate between the two operation modes, hard-switching (HS) and soft-switching (SS). The obtained results provide high accuracy in the detection, making possible the improvement of the converter efficiency with the proposed model.

Keywords Power electronics · Energy management · Electronics · Smart grid · Power efficiency · Artificial intelligence

---

L. A. Fernandez-Serantes (✉) · J.-L. Casteleiro-Roca · J. L. Calvo-Rolle
University of A Coruña, A Coruña, Spain
e-mail: luis.alfonso.fernandez.serantes@udc.es

J.-L. Casteleiro-Roca
e-mail: jose.luis.casteleiro@udc.es

J. L. Calvo-Rolle
e-mail: jose.rolle@udc.es

© The Author(s), under exclusive license to Springer Nature Switzerland AG 2022

G. Marques et al. (eds.), *Machine Learning for Smart Environments/Cities*, Intelligent Systems Reference Library 121,
https://doi.org/10.1007/978-3-030-97516-6_8

8.1 Introduction

Nowadays, environmental care and the concern about the climate change is growing in importance, involving the society and governments [1], but it has not been like this before. Throughout the years the energy generation has varied, from regional power plants to big production areas. In the first steps, the electric energy is generated from renewable sources and at the places where is needed [2].

Later, with the industrialization, the electric energy is generated from fossil fuels as coal, oil or, recently, gas. The production moved outside of the cities or to remote areas where big power plants had been built. Thus, producing greenhouse gasses that contributes with the increase of the earth temperature [2].

With the aim of improving the electric system and towards a green production of energy, the Smart Grid has become one of the trend topics in research. This concept aims to interconnect the demand and consumption at the place where is needed, allowing a bidirectional power flow between the consumers/generators. Moreover, allows a reduction of transport and distribution costs and energy losses, further reduction of the environmental impact while improving the efficiency of the electric system [3].

The different parts in a Smart Grid are interconnected with power converters, where the energy is adjusted to the demand and the power flow is controlled. For example, the power converters adjust the generated electricity from the solar panels to the grid specifications and control the energy that is stored in the battery storage system [4]. They are in charge of regulating and adjusting the demand and consumption in the system.

Moreover, due to the importance of the power electronics in Smart Grid and in power control, the research done in this field has increased over the last years. Their focus is the improvement of efficiency and maximum power delivery of the units, contributing to the best use of energy [5]. The introduction of new materials, as the wide bandgap, or the rapidly development of the Artificial Intelligence is making possible to achieve efficiencies higher than 97–98%, reducing the overall losses in the energy distribution system [6, 7].

This document is structured as follows: first, an introduction to the history of the electric system is done in Sect. 8.2. Then, in Sect. 8.3, the generation and transport of the energy how is done nowadays is described, to follow by the introduction of the Smart Grid concept, Sect. 8.4. Afterwards, the power converters as a key component in the Smart Grid concept are presented in Sect. 8.5 and followed by Sect. 8.6, in which two different switching modes of the power devices is being introduced. Then, the introduction of the Artificial Intelligence in the power electronics is done in Sect. 8.7, proposing a model able to differentiate between the two switching modes. Finally, the results of the model are presented and the conclusions are drawn in Sects. 8.8 and 8.9.

8.2 History of the Electric Systems

The introduction of the electricity happened first as a study tool, where the scientist had tried ways of generation and consumption of the electricity. The electricity in the first times was used for entertainment of the population, with the aim of showing what is capable to do this new discovering [8].

The first years, the scientists and engineers were focusing on how they could use this new energy source, after coal, in the industry. Keeping the focus of how to generate the electric energy to run machines and improve the production of goods. In those ages, end of XIX century, the main use of the electricity was in transmission of information and energy conversion (lightning, motors, etc.). The implementation of the electricity in the industry, gave way to the "Second industrial revolution", substituting the steam machines and with the introduction of electric machines and systems. The electricity introduced flexibility and transportability to the industry, as it can be converted into light, mechanical work or heat and it can be easily transported through cables to any other point where the energy is required [9].

The first steps done in the field of the electric energy were done by Nikola Tesla and by Thomas Edison, were each of them was defending the different types of electricity: alternating current and direct current. While T. Edison had built the first power plant in New York 1892 based on direct current, N. Tesla through the company Westinghouse was illuminating the Chicago's World Fair in 1893 and building in 1896 the hydroelectric power plant at the Niagara waterfalls. At that time, due to the development of electrical machines, like the raise the voltage to reduce conduction losses with a transformer, the development of alternating current generators, etc. the technology proposed by Tesla was chosen, the alternating current [10].

At that time, the electric energy was produced mainly from coal or water, being able to distinguish the pioneers' countries depending on their coal reserve and the abundance of rivers. For example, countries like Finland, Norway, Switzerland or Italy focus on the production of electricity from hydropower plants, as in these countries the rivers flow with high amounts of water, regularly or they have high unevenness of the river´s course. At that time, the power plants were built at a place where the energy was needed [2].

On the other hand, in countries such as Germany, England or United States of America, the main source for producing electric energy was the coal, burning the coal to generate steam and propel a steam turbine generating electricity. In contrast with the hydraulic powerplants that were more disperse in the country territory, the thermal powerplants were built in big generating plants what lead to the construction of distribution and transport power lines. These plants were usually placed in remote places, far from the consumption points. In this way, the countries that focus more on big power plant rather than small hydroelectric plants distributed in different places, had a faster and stronger development of their distribution infrastructure [2].

The next move in the electric system was the introduction of private investments in distribution as well as in the generation. The street illumination was changing from the use of gas to the electricity, and in some countries where the industry needed lot of energy, the electricity as also introduced, for example, in the iron and steel industry. In the latest 1880s, in big cities like Berlin and London, the tram and underground were introduced, using the electric energy as a way of transport. Moreover, the electric motor was substituting the steam machine in the industry [11].

When the production was done locally, the use of dynamos and direct current was more common, distributing the energy with two or three cables, and in case of the companies who chose to generate in alternating current there was no definition of voltage and frequency, and they were mainly used for long distance transport.

In countries with few resources and experience in terms of electrical systems, they received investors from other countries; in case of Portugal, Spain or Russia, they received inversions from countries such as Germany [11].

In the first half of the twentieth century, the main production of electricity was from hydraulic sources, expanding the distribution networks and long-distance transport power lines. During the Worldwide Wars, the price of gas, petrol and coal raised, encouraging the construction of hydraulic power plants. The generation was done in alternating current and using transformers to distribute the energy from the far away located hydraulic power plants to the big cities, avoiding the power plants in the cities and providing an economic price to the population. This system was used during the twentieth century until the 60s, with the introduction of nuclear power and other thermal methods such as gas and petrol.

In 1973 the price of petrol suddenly increased, which made that the electric companies searched for a substituting it by coal or nuclear power. In the last years of the XX century, with the change in the electric market explained in Ref. [12], the generation of electricity moved towards gas turbines and started the introduction of renewable energies.

Since the start of the current twenty-first century, the introduction of renewable energy in the electric system has increased drastically. Every year, new wind parks are built, new solar panel plants are installed and the system moves towards a distributed generation closer to the demand.

8.3 Generation and Transport of Electricity

Traditionally, the generation is done in big power plants situated outside of the cities or in remote places. The generation of the electricity comes from coal, petrol, gas, hydraulic or nuclear power plants. The energy production of these power plants can be adjusted easily to the prognosis of the consumption and the transport of the energy can be adequate in advance to fulfill the demand for energy at any place.

With the aim of delivering the produced electricity at the required place, the countries are full of transports lines. At the beginning of the twentieth century, the lines were short and connecting the closed power plant to the city. Over the years, the power net has been developed and installed all around the countries, first connecting the generation points with the demand and, later, creating interconnection between the different transport lines. Thus, there is a mesh of transport lines that protect the system to any error, providing energy from different points if needed [13].

From the generation, the voltage is raised to a transport high voltage, allowing the transport of electricity with low losses. Usually, these power lines work at a voltage of 400 or 220 kV, used to transmit energy through long distances. When the transport line is close to the delivery point, the voltage is reduced at a substation, place also where the lines are interconnected. Then, the voltage of the lines coming to the cities work at middle voltage, from 50 to 10 kV. Finally, at the place of consumption, the voltage is reduced to 230 V, ready to use by the user. All these systems work on alternating current [13, 14].

Moreover, in some cases when the distances are too long or to interconnect different countries, the voltage is raised higher or converted into direct current, reducing in this way the conduction losses through long distances.

The work of interconnecting the generation with the demand is done by the managers of the network, they are in charge of making an estimation of the demand curve of a country, which approximates quite accurate the demand for the following 24 h. They contact the generation plants to increase or reduce the generation and they connect or disconnect lines at the different substations of the electric system [15].

Nowadays, with the introduction of renewable energies and the high installation of new renewable plants, the generation introduces new variables to the electric system. The main generation from the renewable sources is from:

- Hydraulic: traditionally used and one of the first sources for electricity generation. The water is usually kept in a dam and when the energy is needed, the managers of the network open the dam so that the water flows through the turbines, generating electricity.
- Wind: the energy from the wind is converted into electricity with a windmill. The windmill rotates depending on the velocity of the wind that blows into and generates electricity with the generator installed to the windmill axle.
- Solar plant: a field is filled up with mirrors that point to a tower concentrating the sun rays into a single point. At this point, there is a boiler that heats up with the energy of the sun. The water or other type of liquid becomes steam and moves a turbine, which generates electricity. This type of plant only works during the day and usually are placed in desertic areas.
- Solar panel: in this case, there is no turbine, the sun rays strike into a semiconductor material and the electrons of the material get excited, producing a voltage and a current flow between the two terminals. The generated current is direct and needs to be converted from direct current to alternating current (inverted) to be used by the home appliances.

The installation of these generation types is unavoidable. There is a clear trend of reducing the pollution which affects the energy sector, as most of the energy produced around the world comes from fossil fuels, either for energy generation or transportation [16]. Thus, the different governments around the world have created laws and agreements at the international level in order to reduce pollution, for example, the

Kyoto protocol [17]. Thus, one of the premises on which they are based to reduce the emitted pollutants the renewable energy starts playing a key role [18].

The main drawback of the renewable energy sources is that they might be available when there is no demand, or vice versa, when there is high demand, they are not enough production of energy. To fit the demand and the offer of electric energy with the introduction of renewable sources, the managers of the network need to introduce new models into the prediction systems, that take into account the weather forecasts.

Additionally, the companies are thinking of introducing storage systems to stabilize the power network by storing energy when the demand is low and the offer is high and that they deliver energy on the other way around when the offer is low and the demand is high [19]. Moreover, this storing of energy could become a new business model for some companies. They store the energy when the prices of energy are low, and they resell this stored energy when the prices increase.

8.4 Smart Grid

In the last years, the concept of the smart grid has taken increasing importance and the researchers are investigating how to apply this concept to the actual power network. A Smart grid is a sophisticated system of interconnecting the demand and consumption, allowing bidirectional power flow and different other aspects like availability, controllability, flexibility, reliability, scalability, etc. It also appeared as a way of improving efficiency and reducing the impact of renewable energy, the transport and distribution of electricity [3].

This system contemplates the generation, distribution, and demand of energy locally, although connected to the main electrical grid. The main source of generation is renewable, by placing small power plants around the cities or villages. The generation can be done by solar panels, installing them in the rooftop, walls or even ad some fields beside the houses. Another option would be to place a small wind turbine in the backfield, that generates electricity. This generation of electricity is meant to create systems that are sustainable, that consumes what they generate and avoid big distribution networks. When the generation at any point of the smart grid exceeds the possible consumption of the user, the energy is poured into the grid, in this way supplying the demand in other parts of the city or region. Moreover, when the consumption exceeds the generation, the user is automatically connected to the grid and the energy is delivered from it. In this way, the generation, mainly of renewable origin, takes place at the consumption points, thus avoiding the transport of energy over long distances.

In the smart grid concept, the most difficult and important feature to ensure is that the demand should be equal to the generation. On the other hand, the generation is mainly done by renewable energy which makes it very difficult to predict the generation. For example, the wind may not blow when there is a need to consume electricity, like in the nights, when the difference in temperature causes wind. Also, in the case of the use of solar panels, in winter when there is more need for electricity due to heating systems or boilers for hot water, the production of energy is lower, as the weather is cloudy and the strength of the sun rays is less. Also, when the main production happens, during the day, most of the users are at work, so they cannot consume their production [4].

Even these events when the production is higher than the consumption, something common in power plants based on renewable energies, contributed to destabilizing the system. In this way, the introduction of a smart grid can manage the different energy generation and consumption points.

Moreover, another option to solve this problem, the users can install storage systems. The main use would be to store the produced electricity from renewable sources when the production exceeds the consumption and, vice versa, to deliver electricity when the consumption is higher than the demand. An option could be the use of batteries or fuel cells, by charging or discharging depending on the need. Another strategy related to the store of energy, that can become a business model, is that the user stores energy when the price of the electricity is low, so at this time the user consumes from the power grid, stabilizing the demand with the production and storing the produced energy in the storage system. Then, when the price of the electricity is high, the user delivers the stored energy back to the grid, in this way they deliver the required energy to the grid, with the benefit of stabilizing the grid, the demand and the consumption and earning money from this operation.

With the aim of controlling the energy flow between the different elements in the smart grid, consumers and generators, there is a strong IT network and communication network to interconnect and assure the right functioning of the grid.

In this field, there are many studies going on by researchers all around the world. In Ref. [3] the research tries to forecast the wind in one area to be able to predict the production of energy and, as a complement, in Ref. [4] they research about the critical aspects of introducing wind energy into the electrical network. In terms of the introduction of solar energy into the smart grid system, in Ref. [20] they propose a maximum power point tracking that reacts fast to the solar incident

In general, there are many research studies in this field done around the world: in Ref. [21] they present a smart grid project in Europe; in Ref. [22] how the smart grid system works in China; or in Ref. [23] about the smart grid technologies and applications.

8.5 Power Electronics in Smart Grid

With the aim of interconnecting the generation and the customer, in a traditional power grid, most of the generation and the transport of electricity is carried out mainly in alternating current. The generator needs to synchronize the voltage and frequency with the network and then, through transformers, the voltage is raised to the transport voltage and sent to the grid. On the customer side, the voltage is again reduced to be delivered.

On the other hand, with the inclusion of renewable energies in electricity generation, as in the case of solar panels, the energy is generated in direct current. The voltage generated from the panel is usually boosted up and then is inverted, so converted from direct current into alternating current with a power converter: an inverter. Also, despite the fact that wind turbine machines are, in general, of alternating current, they use power electronics to control and allow greater use of the wind. From the point of view of electric energy transport, it should be noted that, on high voltage lines and long distances, such as underwater lines, in Rio Madeira (Brazil), in Talcher-kola (India), or between Valencia and Balearic Islands, direct current transport is used [14].

Furthermore, regarding electricity consumption, nowadays most of the loads used in a house use direct current, despite being connected to an alternating current network. In each of the household electrical appliances, the alternating current is rectified to obtain direct current, since most electronic circuits work with it.

In a smart grid, the control of the generated, transmitted, or consumed power is carried out by power electronic converters. These converters are in charge of integrating the different types of currents and allow the control of the energy that is transmitted or used. At present, they are used in practically every electrical appliance or electronic device (chargers, power supplies, etc.), as well as in renewable energies, in order to maximize the power generated and transmitted to the electrical network, and in storage systems.

To help the implementation of this type of network and to control the generation, distribution, storage and consumption of electrical energy, power electronic converters are used, among other technologies.

Nowadays, power converters are one of the most widely used electronic devices, since they are the basis of the electrical supply of many electrical and electronic equipment, both for industrial and commercial use. Currently, research is focused on improving current topologies, reducing size, improving efficiency, etc. [6]. Even large international companies are forcing and pushing the development of new technologies and designs through awards in order to improve current devices [24].

The main studies are focused on improving the efficiency of power converters. Through the use of different topologies, control systems, filters used, transistor technologies, etc. the aim is to achieve the reduction of losses [6].

Different topologies are used in order to reduce losses and improve the efficiency of power converters. The first changes appear with the substitution of diodes for controlled transistors, due to the fact that the conduction losses are lower than the diodes, as well as the switching losses due to the reverse recovery effect. There are also different topologies based on resonant circuits that, by certain conditions, the switching is practically lossless, so-called soft-switching (explained in the next section). There are basically two types of conditions: that during the commutation the current is brought to 0 or that the voltage is also worth 0 at that particular moment. For this, the circuits are based on the use of capacitors and coils. As an example, the LLC (inductor-inductor-capacitor) resonant converter [25]. In other cases, it is not possible to achieve any of these conditions, but if a decrease in voltage, coinciding with the moment of switching, they are the so-called quasi-resonant converters, as for example in the flyback type converter [26].

In the last years, the development and research done in the power electronics field have increased with the aim of improving the power converters. Those studies focus on efficiency improvement and size reduction with the introduction of the wide-band bap materials into this field. The materials Silicon Carbide (SiC) and Gallium Nitride (GaN) are already replacing traditional silicon transistors and diodes [27], and with the improvement of the production with these materials, they are becoming more competitive. Moreover, with the wide band-gap materials, the use of different switching techniques and operating modes need to be rethought as they provide better characteristics and can be attractive for many applications [5].

Furthermore, Artificial Intelligence (AI) is expanding rapidly in the research areas, developing systems with intelligence capable of learning as humans. The field of power electronics is also taking the influence of AI, providing solutions in many applications [7, 28]. In the design of components, like in Ref. [28], the AI is used in the design of inductors, improving traditional methods of development and providing accurate results compared with the measured prototypes. In Ref. [7] the algorithms are used to extract the maximum energy from renewable energy systems such as wind turbines or solar panels, tracking the maximum power point in real-time. Most of the machine learning algorithms are based on Multilayer Perceptron as it is an efficient algorithm that delivers feasible and highly accurate predictions. AI has a big potential as it can be used by engineers to improve the traditional solutions and provide a new perspective, providing the best solution, reducing development times and increasing the performance of the designs.

8.6 Hard-Switching and Soft-Switching

In power converters, the transistors are switched between on- and off-state. This transition happens as follows: the turn-on transitions starts when a voltage signal is applied to the gate of the transistor. At this instant, the charges from the gate start building a conductive channel in the device, allowing the current to flow through it; the current rises and the voltage drops. Once the transition is finished and the transistor is saturated, the device is conducting the whole current while keeping the on-state resistance low.

On the other hand, when the voltage signal at the gate goes low, the channel in the transistor starts building down, closing the channel and increasing the resistance. Thus, the current is blocked and the voltage across the device starts rising and finally blocking the full applied voltage. During the turn-off commutation of the switches, there is still current flowing while the voltage starts rising. A similar situation happens during the turn-on commutation when the current starts increasing while the voltage falls. These two transitions caused the switching losses, which provide an addition to the conduction losses, are the main cause of the losses in the power transistors. This operation mode is the so-called Hard-Switching (HS) mode.

The new researches focus on increasing the efficiency of the converter by finding a way of switching with reduced losses. In Soft-Switching (SS) mode, the switching-on and/or switching-off transitions occur during a certain condition. These conditions are usually Zero Voltage Switching (ZVS) or Zero Current Switching (ZCS). The manner in which the switching transition happens is either with current equal to zero or voltage equal to zero, caused by the resonance of components in the circuit. The solution could be as simple as adding a snubber capacitor or even using the output capacitance of the transistor, discharging the channel before the commutation happens. In other cases, the resonance happens thanks to LC (inductor-capacitor) tanks place in the circuit.

As an example, in the LLC converter, the resonance between the inductance and the capacitors, the current drops to 0 A. At this point, the voltage signal is applied to the gate of the transistor making the device starts switching while the current stays at zero. Once the switching transition has been finished, the current starts rising until its maximum value and resonates again, like a sinusoidal wave. At the second point when the current drops to 0 A, the gate signal is removed so the voltage on the device starts rising as it is turned off.

The SS condition is forced with the aim of decreasing the power losses during the commutation so that the power losses are equal to $P(t) = v(t) \cdot i(t) = v(t) \cdot 0 = 0$ W, can be significantly reduced. Thus, thanks to the switching losses reduction, the switching frequency can also be increased, making the passive components smaller and providing higher efficiency to the power converter.

8.7 Artificial Intelligence in Power Electronics

As mentioned above, Artificial Intelligence is being introduced into the power electronics field. One of the main uses of AI is in the design of components [28] or proposing intelligent control systems [7]. The main use is to optimize the designs and achieve higher performance and efficiency to the systems.

An introduction of the optimization issues and control, from the point of view of energy efficiency, will be done. Moreover, the introduction of Artificial Intelligence is presented as a way of improving the efficiency and development of the power converter.

In this research, an intelligent classification model that detects the operating mode of a power converter has been done. The case of the study is a half-bridge buck converter as shown in Fig. 8.1. The reason why it was chosen this topology against others is that this half-bridge is the base of multiple other topologies, like boost, full-bridge, etc. This topology is integrated by two transistors operating in the following manner: when the high-side switch is on, the low-side switch is off, and vice versa. This operation of the switches produces a square signal that needs to be filtered. An LC filter is used to obtain the average value of the square signal and get a constant voltage at the output.

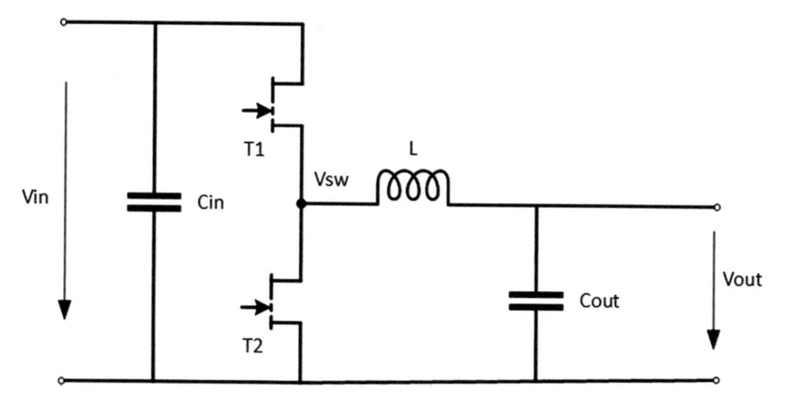

Fig. 8.1 Synchronous buck converter (half-bridge)

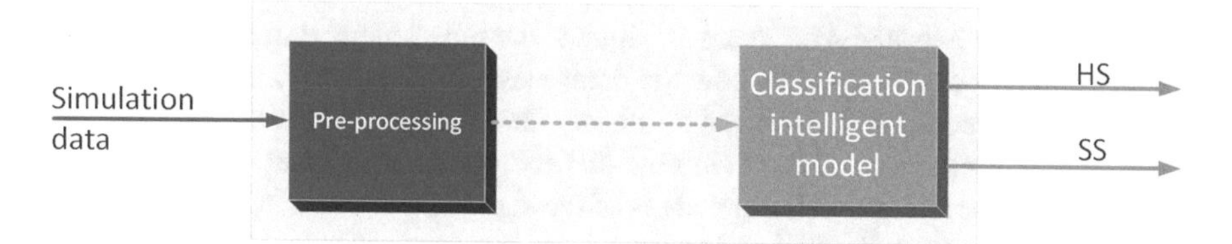

Fig. 8.2 Model approach

In the proposed circuit, the resonance tank is made by the filter inductor and the parasitic output capacitance of the transistors (Coss), used as a non-dissipative snubber [27].

8.7.1 Model Approach

An intelligent classification model has been implemented in this research. This classifier distinguishes between the two possible operating modes of the converter, HS and SS modes. The data used by the intelligent model has been obtained from simulation data. As in Fig. 8.2, this data has been pre-processed to convert the raw data into more representative parameters. Then, with the aim of developing the right model, four different algorithms have been applied.

8.7.2 Dataset

The circuit shown in Fig. 8.1 has been simulated with the LTSpice simulation tool. This circuit has been operated in different points, varying the load at a specific output voltage. With the aim of obtaining consistent data, the circuit keeps unchanged during the whole recompilation of data.

In total, 80 simulation results are used in the dataset, combining both operating modes, HS and SS, 50% of each type. The variables that are measured are the followings:

- Input voltage: the input source is set to 400 V.
- Output voltage: the regulation keeps the voltage at 200 V and filtered to achieve a ripple of 5%.
- Switching node voltage (Vsw node in Fig. 8.1): the voltage jumps from 0 to 400 V, input voltage. The switching frequency varies from 80 kHz to 2 MHz.
- Inductor current: the inductor is charged and discharged. The shape of the current is triangular and in HS the ripple is 20% while in SS the current goes down to 0 A in each cycle.
- Output current: is the current that the load absorb.

Then, this data is analyzed: from all signals, the most interesting is the voltage at the switching node, as it shows how the commutation happens. This voltage signal is used for the model and also used as a base; then, the first and second derivatives are applied with the aim of removing the on- and off-states, as their derivative is 0. Furthermore, the rising and falling edges of the signal, shown in Fig. 8.3 as a dotted line, are separated and the first and second derivatives are applied. Also, the integral of the rising edge data and falling edge data are calculated. Figure 8.3 represents the rising and falling edges of the voltage of the switching node.

As result 8 signals were obtained from the switching node voltage: the raw data, the first and second derivatives of this signal, the rising/falling edge data, the first and second derivatives of the rising/falling edge data, the integral of the rising edge data and, also, the integral of the falling edge data. Moreover, to significantly analyze the data, the following statistics have been calculated for each signal: average, standard deviation, variance, co-variance, root mean square (RMS) and total harmonic distortion (THD), resulting in an 8×6 matrix for each of the 80 simulations.

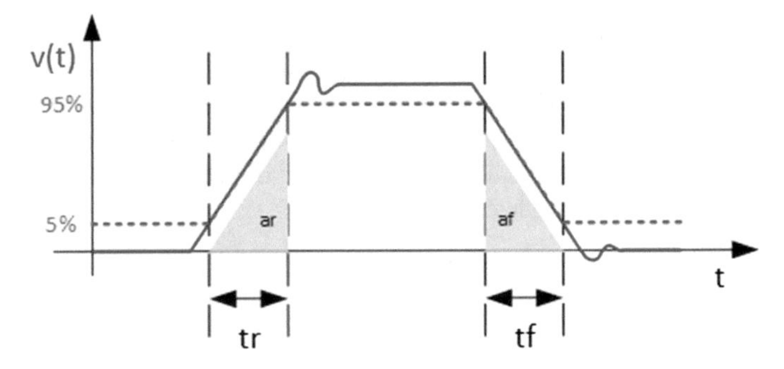

Fig. 8.3 Rising edge and falling edge of the voltage at the switching node used in the model, in dashed blue, and the original signal in continuous red

8.7.3 Methods

The methods used in this work with the aim of improving the classification are Artificial Neural Networks (configured as Multilayer Perceptron, MLP), Support Vector Machines (SVM), the ensemble classifier and, to reduce the dimensionality of the dataset, Linear Discrimination Analysis (LDA). These algorithms are briefly described below.

8.7.3.1 Artificial Neural Networks

The most common ANN (Artificial Neural Network) used is the perceptron. This is a feedforward network with one hidden layer (if it has more, the name would be MultiLayer Perceptron) with a varied number of neurons. It also has an input layer, with the same number of neurons as inputs to the model, and the output layer, which has as many neurons as outputs of the model. Al the neurons in the same layer have the same inputs and outputs, and the same activation function. However, the connections of the inputs have different weights for each neuron. During the training process, these weights are adjusted to minimize the error in the model output [7].

In this chapter, the MLP is configured with only one hidden layer, and there were tested several models with a different number of neurons in the hidden layer: from 1 to 10. As the models are used for classification, the output is a discrete value to assign the input sample to group. There was used Levenberg–Marquardt as a training function.

8.7.3.2 Support Vector Machine

Another commonly used method for classification is the Support Vector Machines. This algorithm uses a kernel operator to increase the dimensionality in the dataset and translate the original samples to a new hyperspace. In this new hyperspace, the algorithm tries to look at two hyperplanes that divide the groups with the maximum distance between them [29].

8.7.3.3 Ensemble

Ensembles used multiple methods to classify the dataset, and this combination improves the performance compared with a single classification algorithm. The process chooses fewer weak learners to perform a regularization and decrease the classification error [30].

8.7.3.4 Linear Discrimination Analysis

The LDA algorithm used in classification involves another translation in the dataset like the SVM, but in this case, the projection is made to reduce the dimensionality of the data. This algorithm provides the projection and the label of the class for each sample. The output is calculated by the analysis that converts a gradual decision to a binary one [31].

8.7.3.5 Classification Performance

With the aim of verifying the performance of the model, validation data is used. First, the model is obtained and it performs the classification with the validation data; then the obtained results are compared with the already known results to verify if the classification has been done correctly. The confusión matrix is the method used to measure the performance of the classification.

A common method for verifying the performance of the classification is the confusion matrix. The matrix is divided as follows: in the columns, the true values are placed while in the rows, the predictions from each model are entered [32]. Then, for each of the predictions obtained by the classifier, the value is entered into the matrix: there are two decision classes, either positive (P) or negative (N). The entries to the confusion matrix are true positives (TP), false positives (FP), true negatives (TN) or false negatives (FN) [32].

From the confusión matrix, 5 different indicators can be extracted to evaluate the performance of the classification algorithm, making easier the comparison of the results. These statistics are the following: sensitivity (SE), specificity (SPC), positive prediction value (PPV), negative prediction value (NPV), and accuracy (ACC) [32].

8.7.4 Experiments Description

The experiments done are carried out in the following manner: first, the data is divided in two groups, one group is the training data, with which the models are trained. The second group of data is used to validate the performance of the obtained models. In this work, groups divide the data in 75% for training and 25% for validation, and the division is done in a random manner. The following algorithms are trained to obtain the different models:

- MLP: the type of algorithm used for the MLP is the Levenberg–Marquardt backpropagation. The number of input and output layers keep constant while the number of neurons on the hidden layers vary from 1 to 10 neurons.
- LDA: the used algorithm provides the same covariance matrix for all classes, meaning that it is regularized.

- SVM: it was trained with a linear kernel function. This function is usually for two-class learning. The used predictors are standard and the predictors are centered and scaled by corresponding weighted column mean and standard deviation.
- Ensemble: the adaptive logistic regression is used for binary classification and the decision tree is used for the weak learners. The cycles used for the ensemble vary from 10 to 100 in steps of 10.

Once the different models are obtained for each technique, the validation is done. For this purpose, the validation data is used and the prediction obtained by the model is compared with it. As result, a confusion matrix is obtained for each of the models and the parameters are calculated.

8.8 Results

In this section, the results obtained by the different models are presented. Table 8.1 summarizes the results obtained by each model and presents the different validation parameters obtained from the confusion matrixes.

As a comparison, the best results have been achieved by the MLP7, which achieves a sensitivity of 0.944, a negative predictive value of 0.941 and, finally, an accuracy of 0.970.

As also shown in Table 8.1, the algorithm that provides better performance is the MLP. Then, the methods of LDA and SVM provides a medium accuracy, 0.602 and 0.779, respectively. Thus, the worse accuracy and classification results are provided by the ensemble, with an accuracy of 0.514.

Table 8.1 Summary of the results

	SE	SPC	NPV	PPV	ACC
MLP1	0.727	1	0.636	1	0.815
MLP2	0.653	1	0.470	1	0.735
MLP3	0.634	0	0	0.970	0.622
MLP4	0.531	1	0.117	1	0.558
MLP5	0.708	1	0.588	1	0.794
MLP6	0.557	1	0.206	1	0.602
MLP7	0.944	1	0.941	1	0.970
MLP8	0.654	1	0.470	1	0.735
MLP9	0.791	1	0.735	1	0.867
MLP10	0.515	1	0.059	1	0.529
SVM	0.694	1	0.559	1	0.779
LDA	0.557	1	0.206	1	0.602
Ensemble 10–100	0.507	1	0.029	1	0.514

The main objective of Smart cities is the efficiency and the optimal use of the energy in the system, reducing the power losses at each part of the energy system. The power converters are a crucial element in the Smart Grid as a single-family house can be both generator or costumer of electricity, the energy distribution and power conversion need to be done as optimal as possible, achieving the highest efficiency. Moreover, the reduction of the converter power losses leads to a benefit for the environment, as the losses don´t need to be covered by the generation source.

The power converters are present in all daily used devices, from a phone, house appliances to the vehicle we drive to the office. The improvement of efficiency becomes a very important topic with the aim of helping the environment. The introduction of machine learning and AI in this field helps the optimization of the switching converter, allowing the reduction of power losses. The proposed system detects and classifies the operation mode of the presented power converter and it could be used for optimizing the efficiency and loss reduction.

8.9 Conclusion

In this chapter, an introduction into the history of the European electric system has been done followed by the description of the smart grid concept and how power electronics take an important role in the transition. Finally, a novel method for efficiency improvement of the power converter based on AI has been introduced. This novel method aims to detect the operating mode of the power converters. In this case, the research has been focused on a synchronized buck converter. The proposed algorithm classifies and predicts the operating model of the buck converter, distinguishing between HS and SS. The data used is obtained from the simulation of the converter using real models of the transistors. There are 5 main signals which are used in the model and, as result, the performance obtained by each algorithm is presented. The MLP with 7 neurons in the hidden layer provides the best performance achieving a 97% of accuracy in the classification. These proposed methods can be very useful in the improvement of existing power converter topologies and, therefore, increasing the efficiency of the smart grid connections.

The topic presented in this chapter is not usual, as two different disciplines are joined to provide a solution: the use of AI to classify and obtain the maximum efficiency out of a power converter. This research allows the autonomous detection of the operating point; presenting an innovative manner of solving a basic and specific problem with a methodology that can be applied to any type of converter. With the aim of continuing this research, the following future works are presented. The first work to do would be the improvement of the obtained results by developing a hybrid intelligent model and reducing the computational costs by applying a dimensional reduction method. Afterwards, the model will be applied to real measured data from a circuit and implement a control loop to keep the converter in SS mode. Moreover, applying to different topologies of power converters converter.

References

1. Marsh, W.M.: Landscape Planning: Environmental Applications, 4th ed. Wiley (2005)
2. Giannetti, R.: La conquista della forza: risorse, tecnologia ed economia nell'industria elettrica Italiana (1883–1940). Angeli (1985)
3. Colak, I., Bayindir, R., Sagiroglu, S.: The effects of the smart grid system on the national grids. In: 2020 8th International Conference on Smart Grid (icSmartGrid), pp. 122–126 (2020). https://doi.org/10.1109/icSmartGrid49881.2020.9144891
4. Lu, S., Sun, K., Cao, G., Yi, Z., Liu, H., Li, Y.: A high efficiency and high power SiC DC-DC converter based on interleaved-boost and full-bridge LLC integration for PV applications. In: 2020 IEEE Energy Conversion Congress and Exposition (ECCE), Detroit, MI, USA, pp. 4822–4827 (2020). https://doi.org/10.1109/ECCE44975.2020.9236122
5. Neumayr, D., Bortis, D., Kolar, J.W.: The essence of the little box challenge-part A: key design challenges & solutions. CPSS Trans. Power Electron. Appl. **5**(2), 158–179 (2020)
6. Colak, I., Fulli, G., Bayhan, S., Chondrogiannis, S., Demirbas, S.: Critical aspects of wind energy systems in smart grid applications. Renew. Sustain. Energy Rev. **52**, 155–171 (2015)
7. Tahiliani, S., Sreeni, S., Moorthy, C.B.: A multilayer perceptron approach to track maximum power in wind power generation systems. In: TENCON 2019–2019 IEEE Region 10 Conference (TENCON), pp. 587–591. IEEE (2019)
8. Aracil, J.: Fundamentos, método e historia de la Ingeniería: una mirada al mundo de los Ingenieros. Síntesis (2010)
9. de Motes, J.M.: Los pioneros de la segunda revolución industrial en España: la Sociedad Española de Electricidad (1881–1894). Rev. Hist. Ind. 121–142 (1992)
10. Manubens, J.C.A., Miñana, J.S.: La introducción de la técnica eléctrica. In: Técnica e ingeniería en España, pp. 649–696 (2011)
11. Hausman, W.J., Hertner, P., Wilkins, M., Electrification, G.: Multinational Enterprise and International Finance in the History of Light and Power, 1878–2007. Cambridge University Press, New York (2008). xxiv+ 487 pp. $80 (hardcover), ISBN: 978-0-521-88035-0
12. Joskow, P.L.: California's electricity crisis. Oxford Rev. Econ. Policy **17**(3), 365–388 (2001)
13. Zhang, J.: Power electronics in future electrical power grids. In: 2013 4th IEEE International Symposium on Power Electronics for Distributed Generation Systems (PEDG), pp. 1–3. IEEE (2013)
14. Paez, J.D., Frey, D., Maneiro, J., Bacha, S., Dworakowski, P.: Overview of DC–DC converters dedicated to HVdc grids. IEEE Trans. Power Delivery **34**(1), 119–128 (2018)
15. Perez, J.A.: Impacto en regulación frecuencia-potencia de los cambios de programa en escalón de las unidades generadoras. In: Anales de Mecánica y Electricidad (2011)
16. Montero-Sousa, J.A., Casteleiro-Roca, J.L., Calvo-Rolle, J.L.: The electricity sector since its inception until the second world war. DYNA **92**, 43–47 (2017)
17. United Nations: Kyoto protocol refence manual on accounting of emissions and assigned amount (2008). https://unfccc.int/sites/default/files/08_unfccc_kp_ref_manual.pdf. Accessed 04 Feb 2021
18. Karunathilake, H., Hewage, K., Mérida, W., Sadiq, R.: Renewable energy selection for net-zero energy communities: life cycle based decision making under uncertainty. Renew. Energy **130**, 558–573 (2019)
19. Kaltschmitt, M., Streicher, W., Wiese, A.: Renewable Energy: Technology, Economics and Environment. Springer-Verlag, Berlin, Heidelberg (2010)
20. Belkaid, A., Colak, I., Isik, O.: Photovoltaic maximum power point tracking under fast varying of solar radiation. Appl. Energy **179**, 523–530 (2016)
21. Colak, I., Fulli, G., Sagiroglu, S., Yesilbudak, M., Covrig, C.F.: Smart grid projects in Europe: current status, maturity and future scenarios. Appl. Energy **152**, 58–70 (2015)
22. Jin, X., Zhang, Y., Wang, X.: Strategy and coordinated development of strong and smart grid. In: IEEE PES Innovative Smart Grid Technologies, pp. 1–4 (2012). https://doi.org/10.1109/ISGT-Asia.2012.6303208

23. Bayindir, R., Colak, I., Fulli, G., Demirtas, K.: Smart grid technologies and applications. Renew. Sustain. Energy Rev. **66**, 499–516 (2016)
24. Gendron-Hansen, A., et al.: High-performance 700 V SiC MOSFETs for the industrial market. In: 2019 IEEE 7th Workshop on Wide Bandgap Power Devices and Applications (WiPDA), pp. 410–415. Raleigh, NC, USA (2019). https://doi.org/10.1109/WiPDA46397.2019.8998881
25. Rahman, A.N., Lee, C., Chiu, H., Hsieh, Y.: Bidirectional three-phase LLC resonant converter. In: 2018 IEEE Transportation Electrification Conference and Expo, Asia-Pacific (ITEC Asia-Pacific), pp. 1–5. Bangkok (2021). https://doi.org/10.1109/ITEC-AP.2018.8433271
26. Xu, S., Shen, W., Qian, Q., Zhu, J., Sun, W., Li, H.: An efficiency optimization method for a high frequency quasi-ZVS controlled resonant flyback converter. In: 2019 IEEE Applied Power Electronics Conference and Exposition (APEC), pp. 2957–2961. Anaheim, CA, USA (2019). https://doi.org/10.1109/APEC.2019.8722026
27. Fernandez-Serantes, L.A., Berger, H., Stocksreiter, W., Weis, G.: Ultra-high frequent switching with GaN-HEMTs using the coss-capacitances as non-dissipative snubbers. In: PCIM Europe 2016; International Exhibition and Conference for Power Electronics, Intelligent Motion, Renewable Energy and Energy Management, pp. 1–8. VDE (2016)
28. Liu, T., Zhang, W., Yu, Z.: Modeling of spiral inductors using artificial neural network. In: Proceedings 2005 IEEE International Joint Conference on Neural Networks, 2005, vol. 4, pp. 2353–2358. IEEE (2005)
29. Mohan, L., Pant, J., Suyal, P., Kumar, A.: Support vector machine accuracy improvement with classification. In: 2020 12th International Conference on Computational Intelligence and Communication Networks (CICN), pp. 477–481 (2020). https://doi.org/10.1109/CICN49253.2020.9242572
30. Uysal, I., Güvenir, H.A.: An overview of regression techniques for knowledge discovery. Knowl. Eng. Rev. **14**(4), 319–340 (1999)
31. Guo, S., Tracey, H.: Discriminant analysis for radar signal classification. IEEE Trans. Aerosp. Electron. Syst. **56**(4), 3134–3148 (2020). https://doi.org/10.1109/TAES.2020.2965787
32. Calvo-Rolle, J.L., Corchado, E.: A bio-inspired knowledge system for improving combined cycle plant control tuning. Neurocomputing **126**, 95–105 (2014)

Chapter 9
Intelligent Simulation and Emulation Platform for Energy Management in Buildings and Microgrids

Tiago Pinto, Luis Gomes, Pedro Faria, Zita Vale, Nuno Teixeira, and Daniel Ramos

Abstract Recent commitments and consequent advances towards an effective energy transition are resulting in promising solutions but also bringing out significant new challenges. Models for energy management at the building and microgrid level are benefiting from new findings in distinct areas such as the internet of things or machine learning. However, the interaction and complementarity between such physical and virtual environments need to be validated and enhanced through dedicated platforms. This chapter presents the Multi-Agent based Real-Time Infrastructure for Energy (MARTINE), which provides a platform that enables a combined assessment of multiple components, including physical components of buildings and microgrids, emulation capabilities, multi-agent and real-time simulation, and intelligent decision support models and services based on machine learning approaches. Besides enabling the study and management of energy resources considering both the physical and virtual layers, MARTINE also provides the means for a continuous improvement of the synergies between the Internet of Things and machine learning solutions.

Keywords Demand response · Internet of things · Machine learning · Multi-agent systems · Real-time simulation · Smart grids

9.1 Introduction

Consumers will play a central role in future Power and Energy Systems (PES), and particularly in Electricity Markets (EM), according to the European Commission's regulation on European Union (EU) internal EM [1]. Consumers' flexibility and active participation in the energy transition and in EM transactions is seen as critical to balance the increasing variability from the generation side. The penetration of Renewable Energy Sources (RES) is being boosted by the ambitious goals and initiatives to increase RES generation and use that are being set all around the globe, including a strong commitment by the EU [2].

T. Pinto (✉) · L. Gomes · P. Faria · Z. Vale · N. Teixeira · D. Ramos
GECAD Research Group, School of Engineering, Polytechnic of Porto, Porto, Portugal
e-mail: tcp@isep.ipp.pt

© The Author(s), under exclusive license to Springer Nature Switzerland AG 2022

G. Marques et al. (eds.), *Machine Learning for Smart Environments/Cities*, Intelligent Systems Reference Library 121,
https://doi.org/10.1007/978-3-030-97516-6_9

Although essential to mitigate climate change concerns and to overcome the problems associated with the scarcity of fossil fuels; the increasing penetration of RES is bringing significant challenges to PES. Multiple models and solutions are being proposed to overcome some of the most significant problems in this domain, e.g. for energy resources management [3], grid planning [4], demand response [5], building energy management [6], novel market models [7], among many others. These solutions, however, are usually presented to solve specific and independent problems, often disregarding the impact that these smaller problems have on each other.

The objective of this chapter is, therefore, to present a solution that enables studying the problem as a whole. The MARTINE (Multi-Agent based Real-Time Infrastructure for Energy) platform has been developed with this purpose [8]. MARTINE incorporates multiple models and solutions including the interaction with real buildings and the respective energy resources, making use of the Internet of Things (IoT) concept. MARTINE is supported by a large set of intelligent models, based on some of the most relevant artificial intelligence technologies, such as multi-agent systems and machine learning, to support the involved players' decisions and real-time control of the physical and virtual resources.

After this introductory section, Sect. 9.2 presents a review on the role of machine learning in IoT, as means to motivate the decisions undertaken when developing MARTINE. The MARTINE platform is presented in Sect. 9.3, including the cyber-physical infrastructure and knowledge layer, where IoT and machine learning come together. An illustrative case comprising the use of machine learning for load forecasting in an office building and the respective interaction with the IoT capabilities in this same building is presented in Sect. 9.4. Finally, Sect. 9.5 presents the main conclusions of this work.

9.2 The Role of Machine Learning in IoT

The use of the IoT, namely IoT devices, is growing and spreading across several domains. Worldwide, at least one IoT device is expected to be installed in 216.9 million homes by 2022 [9], The market of IoT, worth USD 235 billion in 2017, is expected to grow to reach USD 520 billion by 2021 [10]. In order to create innovation, and boost IoT technology, machine learning models can be integrated into IoT systems, enabling the emergence of smarter and intelligent solutions. Machine learning can be integrated into IoT to solve several issues or even to bring new possibilities. This section will describe how machine learning can be used in IoT systems, namely for the domains of security, agriculture, healthcare, smart cities, smart buildings, and industry.

Threats and attacks made to IoT infrastructures/systems are growing. IoT systems can suffer from attacks and anomalies that can cause failures. The use of machine learning models can help to predict and mitigate attacks and anomalies in IoT systems. With the use of logistic regression, support vector machines (SVM), decision trees, random forest, and artificial neural networks (ANN) it is possible to detect these anomalies [11]. Still, in the business of securing IoT systems, machine learning can be used in an IoT gateway to help secure the entire system. In Ref. [12], it is proposed the use of ANN in gateways to enable the detection of anomalies in data sent from edge devices. IoT along with machine learning can be used for authentication, access control, secure offloading, and malware detection to protect data privacy, but some challenges need to be kept in mind when trying to implement certain machine learning-based security techniques, such as communication and computation overhead, backup security solutions and partial state observation [13].

Another of the many concerns regarding the security of IoT systems is the existence of unauthorized IoT devices. Through the use of supervised machine learning models, it is possible to detect unauthorized devices, with the analysis of network traffic data to accurately detect these unauthorized IoT devices. The use of multiclass classifiers can be a wise approach for classification accuracy, detection speed, classifiers' transportability, and resilience to cyber-attacks [14].

Machine learning can also be used to detect poisoning attacks, preventing the decrease of the system's overall performance. To identify poisonous data, a methodology using contextual information regarding the origin and transformation of data points can be used, enabling the effective detection and mitigation of the attack [15].

Intrusion detection models for IoT networks to classify traffic flow using deep learning, namely feed-forward neural networks and multi-class classification can be used to early detect intruders in IoT systems [16]. Learning algorithms can also detect spam data, enabling the categorization of data using a spam score [17].

Security is critical in IoT environments, so with a deep learning-based intrusion detection system (DL-IDS) it becomes simpler to detect security threats in these environments. These systems detect denial of service (DoS) attacks, user-to-root (U2R) attacks, probe attacks, and remote-to-local (R2L) attacks [18]. In order to classify intrusion attacks on IoT systems, a solution was proposed in Ref. [19] considering a variant of Feed-forward Neural Networks (FNN) known as Self-normalizing Neural Network (SNN) that was shown to be a more resilient network than FNNs.

It is important to give the user a high level of privacy in their data. The use of a reinforcement learning (RL)-based privacy-aware offloading scheme can help IoT devices, in the healthcare domain, to protect the user's location and usage patterns [20]. This scheme enables the IoT device to choose an offloading rate that increases computing performance, protects the user's privacy, and lowers the IoT device's consumption. The RL-based offloading scheme, proposed in Ref. [20], uses transfer learning to reduce random exploration at the initial level of the learning process. To be able to have broad IoT solutions, it is needed to integrate/filter/process the data coming from IoT devices to automize the analysis and evaluation of data. This has been a significant problem for developers, due to the size and heterogeneous structures of data. Semantic and learning approaches are the key to solve this problem. To overcome this issue, semantic techniques and machine learning algorithms can be applied to process IoT data [21].

The combination of IoT and artificial intelligence can lead us to many other domain applications, such as agriculture. The monitoring system proposed in Ref. [22] enables the identification of changes in grapes to detect diseases in their early stages, using the Hidden Markov Model and an alarm system to the farmer or the expert. Also, with the use of the system, the farmers have the information about the right timing for fertilizers, pesticide application, and irrigation. This system is useful to increase the farmer's profits and to prevent the vineyards from being affected by diseases. There are several solutions already developed for agriculture, such as the one proposed in Ref. [23] where an early warning framework, which gathers information from soil moisture, nitrogen focus, pH estimation, air viscosity, and CO_2 fixation is used to send warning messages to farmers, using a central server to analyze the data.

Agriculture can be improved through machine learning methods by collecting soil parameter information from the field, using a wireless sensor network (WSN) to collect all the data, and uploading it to the cloud. In Ref. [24], this data is fed to a Long Short-Term Memory (LSTM) network to predict the most suitable crop for the next crop rotation, and the results are sent via an SMS service to the user. To increase food production in smart agriculture, an IoT system based on deep reinforcement learning can be used, in Ref. [25], it is proposed a deep reinforcement learning model to make intelligent decisions, such as determining the amount of water for irrigation to improve the growing environment of crops.

The use of IoT devices together with artificial intelligence can also bring benefits in the healthcare domain. Machine learning can be used to predict the patient's condition and IoT can be used to monitor the patient and communicate to the patient about his stress condition [26]. To address skin cancer, which is one of the most common diseases in the world, de A. Rodrigues et al. [27] proposed a model with twelve CNN with seven different classifier configurations to detect common nevi, atypical nevi, and melanomas, enabling the assistant of doctors in finding skin lesions. Addressing heart diseases, Devi and Kalaivani [28] proposed an IoT-enabled electrocardiogram (ECG) monitoring system able to find the RR interval of the ECG signal to capture the variability of the cardiac rhythm, the proposed solution uses a classification process to classify the disease, enabling the detection of cardiac arrhythmias using a timely diagnosis. Real-time patient monitoring using IoT devices can have delays caused by the communication of data to the cloud. To overcome these delays, Kavitha and Ravikumar [29] proposed a solution with four modules: IoT, data pre-processing, context-aware, and decision-making. This solution uses IoT sensors that have been developed to monitor, track, and sense the activities of older people while providing alarms and notifications sent to doctors with the entire condition of the patient with a very low response time.

Another domain where IoT applications can benefit from machine learning models is smart cities where the volume of data is ideal for machine learning models. Intelligent transportation systems can be improved through the use of IoT and machine learning, focusing on the following categories [30]: route optimization, parking, lights, accident detection/prevention, road anomalies, and infrastructure. Within the smart city domain, anomalies and attacks are also on the rise. In Ref. [31], a Network

Intrusion Detection System (NIDS)-based approach called AD-IoT for detecting IoT attacks on a distributed fog layer is proposed. This solution is able to detect malicious behaviors by evaluating a dataset to detect categorized binary classification before distributing it to fog nodes.

With the increasing popularity of IoT, implementing smart controls in buildings to reduce energy consumption is on the rise. In Ref. [32], a plug-and-play learning framework is proposed where the thermal model of each thermal zone in a building is automatically identified without manual configuration, and a thermal model is able to learn from temperature readings measured by IoT-based smart thermostats. This promising solution is able to provide reliable thermal models for future smart building climate controls. Also, for smart buildings, in Ref. [33], it is proposed a real-time system able to detect the user's indoor location and to detect the context of their environment. The system uses sensing and positioning data to classify, using a weighted k-nearest neighbor model, the type of indoor environment around the users. In Ref. [34], it is proposed a semi-supervised deep reinforcement learning model to solve the problem of indoor location by using the intensity of the Bluetooth low energy (BLE) signals. The use of reinforcement learning is a good approach for IoT applications since it requires little supervision for giving feedback through rewards as it learns to choose the best policy among many alternative actions.

The consumption of untreated water causes around 3.4 million deaths per year, but with the use of a low-cost water quality monitoring system using IoT, machine learning, and cloud computing, the traditional water quality monitoring systems can be improved and allow mitigation actions [35]. In Ref. [36], it is proposed a deep learning model to optimize performance in big biological data systems supported by IoT devices.

In Ref. [37], a merge between IoT and machine learning is proposed for supporting museums to design new strategies and operational decisions. The proposed solution collects data regarding visiting paths using a non-invasive way and infers knowledge using clustering models to discover people's behaviors.

In the industrial domain, IoT and machine learning, enable the prediction of remaining useful life (RUL) times to support the maintenance of industrial machines. In Ref. [38], it is proposed an Autoregressive Integrated Moving Average (ARIMA) forecasting model that receives data from several sensors to predict the failures and quality defects. The use of machine learning proves to be a vital component in IoT, as it decreases the cost of maintenance and increases the overall manufacturing process efficiency [38].

IoT devices are limited in computing and communication resources, which are hindering the development of intelligent solutions using machine learning models. However, currently, it is possible to integrate deep learning models into IoT devices, by compiling deep learning models into executable code for a target platform/hardware. One benefit of this integration is that no matter how complicated the models get, you can easily implement them on IoT devices, creating a good solution for designing smarter IoT devices [39].

Through frameworks for machine learning and IoT knowledge discovery, new opportunities arise leveraging the full power of machine learning in IoT systems, providing value and benefits to users [40]. IoT systems can be difficult to understand because of the amount of data it produces, but through the use of unsupervised learning techniques applied, it is possible to support decision making processes within intelligent environments. Unsupervised learning models are able to detect hidden behaviors and context-sensitive situations using in-depth analysis of data [37].

9.3 MARTINE—Multi-agent Based Real-Time Infrastructure for Energy

MARTINE is a Multi-Agent based Real-Time Infrastructure for Energy that is being continuously developed at the GECAD/ISEP campus with the goal of supporting, managing and studying energy management related decisions [8]. As shown by Fig. 9.1, MARTINE is composed of several layers, which comprise the effective management and control of real buildings and their respective resources; the real-time simulation of large scale problems, including the direct interaction with the physical resources and the emulation of resources that are not physically present in the laboratorial infrastructure. Complementarily, and in order to extend the platform capabilities with added flexibility, a multi-agent system layer is used. This multi-agent layer enables, on one hand, representing and modeling players that cannot be physically present (or emulated) in the platform (e.g. a specific type of aggregator, or a wholesale market operator, among many others), while also enabling the distributed control and decision-making of the several involved players. On top of all these models there is a knowledge layer, which integrates the necessary machine learning, optimization and modelling algorithms and services that are used for supporting players' decision-making. MARTINE's cyber-physical infrastructure and the knowledge layer, including its interaction with the multi-agent models are presented in the following sub-sections.

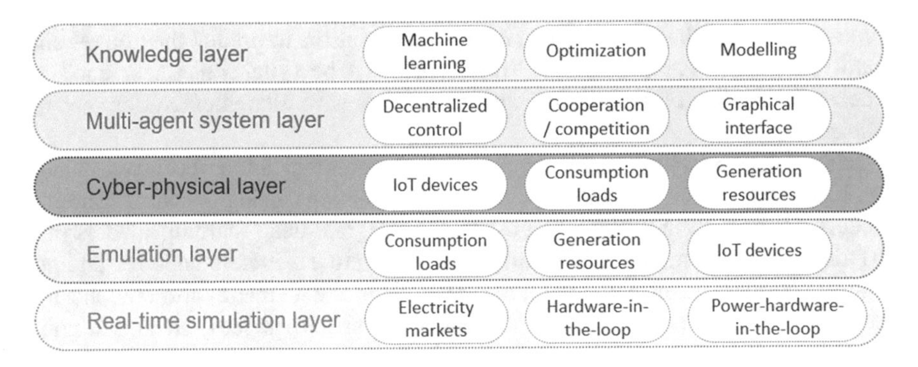

Fig. 9.1 MARTINE's architectural layers

9.3.1 Cyber-Physical Infrastructure

The cyber-physical infrastructure is a differentiation aspect in MARTINE. This layer allows the integration of physical resources in the simulation environment provided by MARTINE architecture. The cyber-physical is then responsible for the integration of real resources usually deployed in uncontrollable environments. Although the integration of real resources demands real-time simulations, it also enables the use of real case studies in a combined environment of simulation, emulation, and reality.

Keeping with the growth of IoT devices, MARTINE enables the integration of market available and custom-made/'do it yourself' (DIY) IoT devices to provide data to its simulations. For instance, it is possible to integrate smart plugs with energy metering and temperature sensors in MARTINE to control physical heating systems according to electricity market simulator prices and the actual temperate measured by the IoT device. The integration of IoT is done using direct control and monitoring over the IoT device, without the need for a cloud system. However, MARTINE also enables the integration of IoT devices through cloud solutions, such as the integration of IoT devices using the webbooks of the IFTTT (If This Then That) cloud platform.

To enable the integration of physical devices into MARTINE, it is available a set of connection drivers. The integration is based on the IEEE STD 2660.1, allowing tightly coupled and loosely coupled integrations. The available drivers implemented, so far, in MARTINE are RS485, Modbus/RTU, DALI (Digital Addressable Lighting Interface), Modbus/TCP, REST-based HTTP requests, MQTT (Message Queuing Telemetry Transport), and AMQP (Advanced Message Queuing Protocol). However, it is possible to develop new drivers at any time to enable more protocols.

The integration of market available IoT devices can be difficult because of the existence of closed protocols. To minimize the impossibility of direct integration, third-party solutions can be used, such as the already mentioned IFTTT. However, local deployable solutions, especially open-source, are more suitable, such as the solutions provided by Home Assistant and openHAB. These open-source solutions allow the integration of market available IoT devices and provide an API (Application Programming Interface) that can be used to monitor and control the IoT devices integrated into those solutions. This allows the use of market available IoT devices by custom-made platforms, such as MARTINE.

The combination of simulation, emulation and physical resources creates a powerful platform. However, due to the particularities of physical resources, security measurements must be considered to prevent the harming and damage of physical resources. When connecting physical resources, the simulation must run in real-time as physical resources, especially sensors, can only operate in real-time. When creating a cyber-physical system, it is important to prevent the continuous sending of signals to/from IoT devices as their limited computation power cannot be enough to correctly operate with high-frequency communication. Also, it is very important to prevent the continuous sending of control signals to physical resources. For instance, if the cyber part of the system is optimization the intensity of a physical lamp, the sending of several signals per second can damage the lamp and negatively impact the use of the illuminated area. Therefore, the use of physical resources must be carefully done in order to prevent damage to resources.

For the integration of physical sensors, the issues of failures and errors in data must also be considered. In MARTINE, there is no default action for these issues, leaving to the user the necessary implementation of such mitigation measures. Code for data imputation and data filtering can be added in the driver class or even on the agent-side.

Besides IoT devices, MARTINE has a significant connection with SCADA (supervisory control and data acquisition) systems, namely in the energy domain. The integration of energy analyzers, invertors, and PLCs (programmable logic controllers) is an important aspect in MARTINE to enable the near real-time energy monitoring and the control of energy-related loads and resources—for emulation and physical control. Therefore, MARTINE has drivers for Modbus, in its two variants TCP/IP (Modbus/TCP) and RS485 (i.e., Modbus/RTU), enabling the direct integration of the major of metering, monitoring, and control units for energy monitoring and control.

9.3.2 Knowledge Layer

MARTINE's knowledge layer includes a large set of models and algorithms that enable supporting the involved players' decisions. The models include forecasting approaches for several resources such as energy generation, consumption, electricity market prices; energy management models for distinct problems, such as building energy management and smart grid energy resources management; and also the simulation and support of players negotiations in electricity markets.

MARTINE platform allows the testing and validation of building energy management strategies. It is, in fact, a relevant topic in the literature as it contributes to the improvement of energy efficiency in buildings, dealing with inhabitants' comfort aspects [41]. Load forecasting can support the building energy management by providing a prediction of the energy consumption in the future, so accurate decisions can be made in advance, allowing also the participation in demand response events with reduced impact on the user comfort. In Ref. [42], the information provided by several sensors in the building is used to support the energy consumption forecast. Regarding energy consumption optimization, Khorram et al. [43] provides an optimization model for the complete building, which includes air conditioning device optimization and participation in demand response opportunities. For a more complete solution regarding the integration of systems, the work in Ref. [44] presents a knowledge-based approach for buildings management.

Moving to the grid level, namely in the smart grid context, several adequate energy resource management methods have been proposed and applied in MARTINE. For the energy management evaluation, a methodology based on key performance indicators has been proposed in Ref. [45]. It is a very important tool to assess the number of benefits obtained from such strategies focusing on the consumer side. Moreover, for the consumer preferences modeling, the work in Ref. [46] provides a methodology for the definition of complex contracts where consumers establish their preferences regarding remuneration and flexibility in the participation in DR programs. Other approaches supported by MARTINE are based on optimization for different aspects in smart grids.

One of the most important aspects when managing energy resources both at the building and smart grid level, is to consider the impact of electricity market participation. In this way, an electricity market simulator, based on a multi-agent approach has been developed an integrated into MARTINE. This simulator is called MASCEM (Multi-Agent Simulator of Competitive Electricity Markets) [47], and supports the simulation of several electricity market models, by modeling the main involved entities through software agents. A decision support system that supports negotiating players in their decisions has also been developed—Adaptive Decision Support for Electricity Markets Negotiations (AiD-EM) [48]. AiD-EM includes decision support features for planning purposes, enabling players to decide in advance in which from a set of available market opportunities to participate. Once the planning is performed, two specific systems are used to support the negotiation process in both bilateral negotiations and auction-based markets. The Adaptive Learning strategic Bidding System (ALBidS) provides a large number of negotiation algorithms for auction-based markets' negotiations and incorporates a reinforcement learning model that enables the system to learn autonomously which are the best strategies to apply in each moment and context. The Decision Support for energy Contracts Negotiations (DECON) system provides the necessary tools to support players' decisions in bilateral negotiations. Specific models for local electricity market negotiations have also been developed, e.g. Pinto et al. [49], enabling players' negotiations at the local communities level, and supporting the interaction with the traditional larger-scale electricity markets.

9.4 Illustrative Case

In this illustrative case study, it presented two distinct examples. The first one concerns load forecasting in an office building. The second one is related to load consumption emulation in a laboratory.

9.4.1 Load Forecasting

The forecasting activities consider a large historic from 22 May 2017 to 17 November 2019 to predict consumption targets of a single week featuring 18 to 24 November 2019. The input structure of train and test data features consumptions placed in periods of five minutes. These consider all the days of the week from Monday to Sunday with no exclusion. The day of the week is discarded from the input structure as the sequence of consumptions featuring independent days of the week already provides an overview of the consumption behavior of each day. Therefore, the adding of the day of the week is additional information that leads to overfitting. Collected data from sensors devices is discarded as well from the input structure including the photovoltaic voltage, temperature, humidity, light intensity, air quality and CO_2. This decision was taken as the large historic of data may result in many weekly patterns where the sensors data has a week correlation with the consumption data. With all this in mind, the input corresponds to ten entries that feature consumption data under five minutes periods that anticipate the targeted consumption. The forecasting algorithms used for these activity tasks correspond to an artificial neural network and k-nearest neighbor. The artificial neural network algorithm considers fixed parameterizations including the learning rate assigned to 0.001, the number of neurons assigned to 64, the clipping ratio assigned to 5, the number of epochs assigned to 500, the early stopping assigned to 20 and the validation split assigned to 0.2. Four scenarios were performed under these conditions and feature similar studies from other publications [42].

The results of the planned scenarios are presented in Table 9.1. This evidences forecasting errors from three possible metrics: WAPE, SMAPE, RMSPE. All these scenarios feature the same conditions including the number of input entries assigned to ten, while these correspond to different executions. Other similar parameterizations for these scenarios are presented in the case study. The forecasting errors are associated with two algorithms including artificial neural networks and k-nearest neighbors.

Table 9.1 shows that the ANN algorithm features less forecasting errors compared to KNN. Additionally, the SMAPE metrics evidence lower forecasting errors under a range between 5 and 9%. The WAPE metrics features higher errors than SMAPE however with a lower difference using ranges between 6 and 10%. Additionally, the RMSPE features higher errors under a range between 11 and 14%. It is further observed that while ANN algorithm features more accurate forecasts, k-nearest neighbors features on all executions lower ranges for each forecasting metrics. The average and the similar observations for the different executions support these insights. The comparison of the real consumptions with the forecasting counterparts is analyzed with more detail in Figs. 9.1 and 9.2 associated respectively to artificial neural networks and k-nearest neighbors.

Table 9.1 Forecast errors

#	Entries	WAPE ANN (%)	SMAPE ANN (%)	RMSPE ANN (%)	WAPE KNN (%)	SMAPE KNN (%)	RMSPE KNN (%)
1	10	6.46	5.23	11.07	7.10	5.63	13.47
2	10	7.39	6.05	11.42	8.19	6.75	13.43
3	10	9.42	8.74	13.68	8.19	6.75	13.43
4	10	7.69	6.54	11.35	8.19	6.75	13.43
	Average	7.74	6.64	11.88	7.92	6.47	13.44

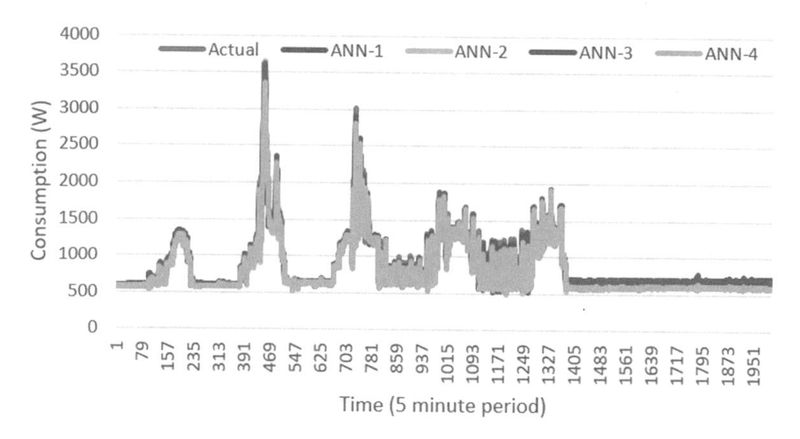

Fig. 9.2 Consumption forecasts from 18 to 24 November 2019 with ANN

The visualization presented in Figs. 9.2 and 9.3 shows that the forecasting activities present very accurate predictions for all five minutes periods both for artificial neural networks and k-nearest neighbors. All the four executions present very similar behavior to the real consumption counterparts being this more truthful for k-nearest neighbors as this algorithm presents almost consistent patterns. These accurate results support the low forecasting errors presented in Table 9.1. More concretely, the consumption presents a sequence of five daily patterns with activity from Monday to Friday followed by the low activity consumption from the weekend. The targeted week from 18 to 24 November 2019 presents a Monday with lower activity compared to the other days reaching a maximum of 1500 W. The consumption activity belonging to Tuesday and Wednesday is higher presenting maximums respectively of 3500 and 3000 W.

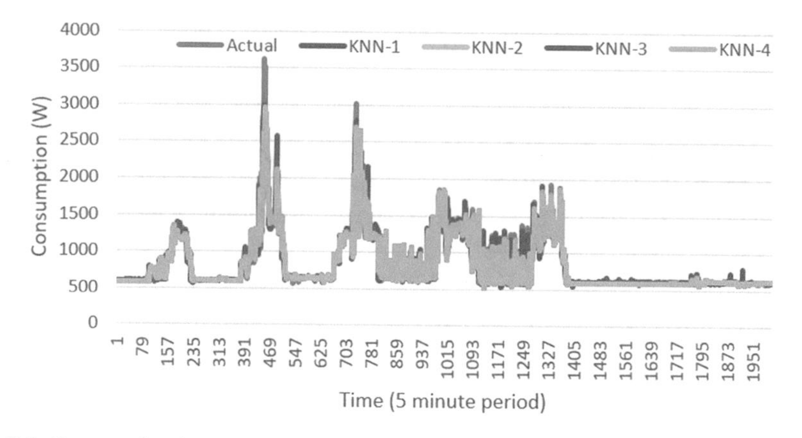

Fig. 9.3 Consumption forecasts from 18 to 24 November 2019 with KNN

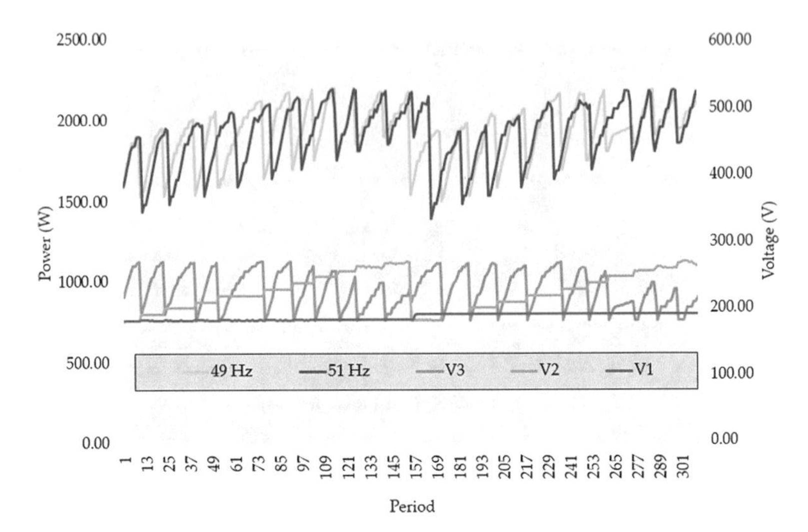

Fig. 9.4 Emulation of a 2.2 kW load

Thursday and Friday present higher consumption than Monday with no passing over 2000 W. The low consumption is presented during the whole weekend being this also the case for the time between the late afternoon and the morning.

9.4.2 Load Consumption Emulation

Another application relevant to the platform description is the emulation of load consumption. Figure 9.4 presents the results of emulation of a 2.2 kW load.

The emulated load is of resistive type. As it can be seen in Fig. 9.4, the three phase voltages have been applied, changing over time, for different ranges of frequencies. Figure 9.4 is only presented results for 49 and 51 Hz in order to make it easy to read. It is possible to conclude that the resistive load consumption is not always 2.2 kW as it is the rated power. In fact, it depends on the voltage applied and on the frequency. In cases where the load manufacturer is not providing sufficient information for the load modeling, the developed work allows a more accurate load modeling.

9.5 Conclusions

This chapter discusses the synergy between IoT and machine learning through a practical and concrete example: MARTINE. MARTINE is a simulation, emulation and intelligent energy management platform for the study of problems related to buildings, smart grids and electricity markets. MARTINE brings together and boosts the complementarity between solutions based on software models for decision support and on physical infrastructure and control approaches. The presented review on the role of machine learning in IoT is used to highlight how the two fields can benefit from each other, motivating the decisions undertaken when developing MARTINE. The several layers that compose MARTINE are described, comprising the effective management and control of real buildings and their respective resources; the real-time simulation of large scale problems, including the direct interaction with the physical resources and the emulation of resources that are not physically present in the laboratorial infrastructure. Complementarily, the multi-agent approach and knowledge layer, which integrates the necessary machine learning, optimization and modelling algorithms and services that are used for supporting players' decision-making are also presented. The illustrative case results demonstrate how a comprehensive platform such as MARTINE can take the most advantage out of the two multidisciplinary fields of IoT and machine learning with practical advantages in a real office building. Such platform thereby offers a relevant environment to experiment with new solutions related to the synergy between IoT and machine learning models, allowing the assessment of the impact that physical and virtual models have on each other using a real smart environment.

Acknowledgements This work is supported by FEDER Funds through COMPETE program and by National Funds through FCT under projects UIDB/00760/2020, UIDP/00760/2020, CEECIND/01811/2017 and CEECIND/02887/2017.

References

1. European Parliament and Council of the EU: Directive (EU) 2019/944 on common rules for the internal market for electricity and amending Directive 2012/27/EU. Off. J. Eur. Union, no. L 158, p. 18 (2019)
2. European Commission: 2030 climate & energy framework—climate action. In: 2030 Climate & Energy Framework, p. 1 (2018)
3. Zhou, B., et al.: Smart home energy management systems: concept, configurations, and scheduling strategies. Renew. Sustain. Energy Rev. **61**, 30–40 (2016)
4. Salehi, J., Haghifam, M.-R.: Long term distribution network planning considering urbanity uncertainties. Int. J. Electr. Power Energy Syst. **42**(1), 321–333 (2012)
5. Xu, X., Chen, C., Zhu, X., Hu, Q.: Promoting acceptance of direct load control programs in the United States: financial incentive versus control option. Energy **147**, 1278–1287 (2018)
6. Afrakhte, H., Bayat, P., Bayat, P.: Energy management system for smart house with multi-sources using PI-CA controller. In: 2016 Iranian Conference on Renewable Energy Distributed Generation (ICREDG), pp. 24–31 (2016)
7. Chen, T., Su, W.: Local energy trading behavior modeling with deep reinforcement learning. IEEE Access **6**, 62806–62814 (2018)
8. Pinto, T., Gomes, L., Faria, P., Sousa, F., Vale, Z.: MARTINE: Multi-agent based real-time infrastructure for energy. In: 19th International Conference on Autonomous Agents and Multiagent Systems (AAMAS 2020), (2020)
9. Statista: Smart Home Report 2018—Control and Connectivity. Hamburg (2018)

10. Columbus, L.: IoT market predicted to double by 2021, reaching $520B. Forbes (2021). [Online]. Available: https://www.forbes.com/sites/louiscolumbus/2018/08/16/iot-market-predicted-to-double-by-2021-reaching-520b
11. Hasan, M., Islam, M.M., Zarif, M.I.I., Hashem, M.A.A.: Attack and anomaly detection in IoT sensors in IoT sites using machine learning approaches. Internet of Things **7**, 100059 (2019)
12. Cañedo, J., Skjellum, A.: Using machine learning to secure IoT systems. In: 2016 14th Annual Conference on Privacy, Security and Trust (PST), pp. 219–222 (2016)
13. Xiao, L., Wan, X., Lu, X., Zhang, Y., Wu, D.: IoT Security techniques based on machine learning: how do IoT devices use AI to enhance security? IEEE Signal Process. Mag. **35**(5), 41–49 (2018)
14. Yair Meidan, Y.E., Bohadana, M., Shabtai, A., Ochoa, M., Tippenhauer, N.O., Guarnizo, J.D.: Detection of unauthorized IoT devices using machine learning techniques. arXiv (2017)
15. Baracaldo, N., Chen, B., Ludwig, H., Safavi, A., Zhang, R.: Detecting poisoning attacks on machine learning in IoT environments. In: 2018 IEEE International Congress on Internet of Things (ICIOT), pp. 57–64 (2018)
16. Ge, M., Fu, X., Syed, N., Baig, Z., Teo, G., Robles-Kelly, A.: Deep learning-based intrusion detection for IoT networks. In: 2019 IEEE 24th Pacific Rim International Symposium on Dependable Computing (PRDC), pp. 256–25609 (2019)
17. Makkar, A., Garg, S., Kumar, N., Hossain, M.S., Ghoneim, A., Alrashoud, M.: An efficient spam detection technique for IoT devices using machine learning. IEEE Trans. Ind. Informatics **17**(2), 903–912 (2021)
18. Otoum, Y., Liu, D., Nayak, A.: DL-IDS: a deep learning–based intrusion detection framework for securing IoT. Trans. Emerg. Telecommun. Technol. e3803 (2019)
19. Ibitoye, O., Shafiq, O., Matrawy, A.: Analyzing Adversarial attacks against deep learning for intrusion detection in IoT networks. arXiv (2019)
20. Min, M., et al.: Learning-based privacy-aware offloading for healthcare IoT with energy harvesting. IEEE Internet Things J. **6**(3), 4307–4316 (2019)
21. Balakrishna, S., Thirumaran, M., Solanki, V.K.: IoT sensor data integration in healthcare using semantics and machine learning approaches. In: Balas, V.E., Solanki, V.K., Kumar, R., Ahad, M.A.R. (eds.) BT—A Handbook of Internet of Things in Biomedical and Cyber Physical System, pp. 275–300. Springer, Cham (2020)
22. Patil, S.S., Thorat, S.A.: Early detection of grapes diseases using machine learning and IoT. In: 2016 Second International Conference on Cognitive Computing and Information Processing (CCIP), pp. 1–5 (2016)
23. Ding, X., Xiong, G., Hu, B., Xie, L., Zhou, S.: Environment monitoring and early warning system of facility agriculture based on heterogeneous wireless networks. In: Proceedings of 2013 IEEE International Conference on Service Operations and Logistics, and Informatics, pp. 307–310 (2013)
24. Varman, S.A.M., Baskaran, A.R., Aravindh, S., Prabhu, E.: Deep learning and IoT for smart agriculture using WSN. In: 2017 IEEE International Conference on Computational Intelligence and Computing Research (ICCIC), pp. 1–6 (2017)
25. Bu, F., Wang, X.: A smart agriculture IoT system based on deep reinforcement learning. Future Gener. Comput. Syst. **99**, 500–507 (2019)
26. Pandey, P.S.: Machine learning and IoT for prediction and detection of stress. In: 2017 17th International Conference on Computational Science and Its Applications (ICCSA), pp. 1–5 (2017)
27. de A. Rodrigues, D., Ivo, R.F., Satapathy, S.C., Wang, S., Hemanth, J., Filho, P.P.R.: A new approach for classification skin lesion based on transfer learning, deep learning, and IoT system. Pattern Recogn. Lett. **136**, 8–15 (2020)
28. Devi, R.L., Kalaivani, V.: Machine learning and IoT-based cardiac arrhythmia diagnosis using statistical and dynamic features of ECG. J. Supercomput. **76**(9), 6533–6544 (2020)
29. Kavitha, D., Ravikumar, S.: IOT and context-aware learning-based optimal neural network model for real-time health monitoring. Trans. Emerg. Telecommun. Technol. **32**(1), e4132 (2021)

30. Zantalis, F., Koulouras, G., Karabetsos, S., Kandris, D.: A review of machine learning and IoT in smart transportation. Future Internet **11**(4) (2019)
31. Alrashdi, I., Alqazzaz, A., Aloufi, E., Alharthi, R., Zohdy, M., Ming, H.: AD-IoT: anomaly detection of IoT cyberattacks in smart city using machine learning. In: 2019 IEEE 9th Annual Computing and Communication Workshop and Conference (CCWC), pp. 305–310 (2019)
32. Zhang, X., Pipattanasomporn, M., Chen, T., Rahman, S.: An IoT-based thermal model learning framework for smart buildings. IEEE Internet Things J. **7**(1), 518–527 (2020)
33. AlHajri, M.I., Ali, N.T., Shubair, R.M.: Classification of indoor environments for IoT applications: a machine learning approach. IEEE Antennas Wirel. Propag. Lett. **17**(12), 2164–2168 (2018)
34. Mohammadi, M., Al-Fuqaha, A., Guizani, M., Oh, J.: Semisupervised deep reinforcement learning in support of IoT and smart city services. IEEE Internet Things J. **5**(2), 624–635 (2018)
35. Koditala, N.K., Pandey, P.S.: Water quality monitoring system using IoT and machine learning. In: 2018 International Conference on Research in Intelligent and Computing in Engineering (RICE), pp. 1–5 (2018)
36. Irshad, O., Khan, M.U.G., Iqbal, R., Basheer, S., Bashir, A.K.: Performance optimization of IoT based biological systems using deep learning. Comput. Commun. **155**, 24–31 (2020)
37. Piccialli, F., Cuomo, S., di Cola, V.S., Casolla, G.: A machine learning approach for IoT cultural data. J. Ambient Intell. Humaniz. Comput. (2019)
38. Kanawaday, A., Sane, A.: Machine learning for predictive maintenance of industrial machines using IoT sensor data. In: 2017 8th IEEE International Conference on Software Engineering and Service Science (ICSESS), pp. 87–90 (2017)
39. Tang, J., Sun, D., Liu, S., Gaudiot, J.: Enabling deep learning on IoT devices. Computer (Long. Beach. Calif.) **50**(10), 92–96 (2017)
40. Adi, E., Anwar, A., Baig, Z., Zeadally, S.: Machine learning and data analytics for the IoT. Neural Comput. Appl. **32**(20), 16205–16233 (2020)
41. Mariano-Hernández, D., Hernández-Callejo, L., Zorita-Lamadrid, A., Duque-Pérez, O., Santos García, F.: A review of strategies for building energy management system: model predictive control, demand side management, optimization, and fault detect & diagnosis. J. Build. Eng. **33**, 101692 (2021)
42. Ramos, D., Teixeira, B., Faria, P., Gomes, L., Abrishambaf, O., Vale, Z.: Using diverse sensors in load forecasting in an office building to support energy management. Energy Rep. **6**, 182–187 (2020)
43. Khorram, M., Faria, P., Abrishambaf, O., Vale, Z.: Consumption optimization in an office building considering flexible loads and user comfort. Sensors **20**(3) (2020)
44. Santos, G., Vale, Z., Faria, P., Gomes, L.: BRICKS: Building's reasoning for intelligent control knowledge-based system. Sustain. Cities Soc. **52**, 101832 (2020)
45. Faria, P., Lezama, F., Vale, Z., Khorram, M.: A methodology for energy key performance indicators analysis. Energy Informatics **4**(1), 6 (2021)
46. Faria, P., Vale, Z.: Distributed energy resources scheduling with demand response complex contract. J. Mod. Power Syst. Clean Energy 1–13 (2020)
47. Santos, G., Pinto, T., Praça, I., Vale, Z.: MASCEM: Optimizing the performance of a multi-agent system. Energy **111**, 513–524 (2016)
48. Pinto, T., Vale, Z.: AiD-EM: Adaptive decision support for electricity markets negotiations. In: Proceedings of the Twenty-Eighth International Joint Conference on Artificial Intelligence, (IJCAI-19), pp. 6563–6565 (2019)
49. Pinto, T., Faia, R., Ghazvini, M.A.F., Soares, J., Corchado, J.M., do Vale, Z.M.A.: Decision support for small players negotiations under a transactive energy framework. IEEE Trans. Power Syst. 1 (2018)

Chapter 10
Machine Learning Applications and Security Analysis in Smart Cities

İsa Avci and Cevat Özarpa

Abstract The development of technologies, smart grids, and cities has also been increasing in recent years. At the beginning of these technologies are the Internet of Things (IoT), artificial intelligence, and machine learning. In addition to these technologies, the connections of systems used in smart cities with IoT technologies gain importance. The basic principle of smart cities is that all systems used are smart, controllable, and remotely manageable. In addition, the sustainability of IoT systems in smart cities and providing added value in terms of cost is an important issue. This chapter, it is aimed to propose solutions by taking precautions in terms of security, the rise in the use of IoT technologies in smart cities and networks. When the applications developed with machine learning in smart cities are examined, smart stalls, smart traffic lights, smart waste, air pollution and quality forecasts, adequate transportation, smart parking, smart hotel, and smart health systems stand out as the most frequently used applications. While smart cities are designed to increase productivity and efficiency, neglecting security poses serious risks. In this chapter, the security risks of machine learning applications in smart cities are evaluated.

Keywords Smart cities · Machine learning · Secure cities · IoT security · SCADA

10.1 Introduction

Smart cities use the Internet of things (IoT) technology to collect data, manage resources and services efficiently, and make meaningful use of the information obtained from these data. For this reason, the provision of these services to the society living in cities has achieved living conditions in quality living conditions. In addition, Information, and Communication Technologies (ICT) aims to increase

İ. Avci (✉)
Computer Engineering, Engineering Faculty, Karabuk University, 78050 Karabuk, Turkey
e-mail: isaavci@karabuk.edu.tr

C. Özarpa
Mechanical Engineering, Engineering Faculty, Karabuk University, 78050 Karabuk, Turkey
e-mail: cevatozarpa@karabuk.edu.tr

© The Author(s), under exclusive license to Springer Nature Switzerland AG 2022

G. Marques et al. (eds.), *Machine Learning for Smart Environments/Cities*,
Intelligent Systems Reference Library 121,
https://doi.org/10.1007/978-3-030-97516-6_10

efficiency in operational processes. Thus, it should ensure to share information with the public, when necessary, to increase the quality of government services and the welfare of citizens.

The main purpose of a smart city is while improve the quality of living conditions for the society by making data analysis with smart and new technologies, the reason for being a city is to optimize the functions that need to be done and contribute to economic growth. Several main features are used to determine the smartness of a city. These features include [1]:

- It should have a technology-based infrastructure,
- It should contribute to the environment as environmental initiatives,
- There should be a high-capacity and functional public transportation system,
- There should be a quality and sustainable urban planning approach.

The development of information and communication technologies contributes to the development of smart cities. For cities to be smart, technologies must be used. In addition to the increasing population in big cities, the need for smart networks comes to the fore due to the increase in construction investments and megacities in energy demand. The smart grid serves many areas. The main ones are smart health, smart transportation, smart wholesale, smart vehicles, smart security, and smart meters. The development of all fields shows change and development primarily depending on the development of technology. Along with smart grids, the concept of a smart city is critical. Optimum operation of smart applications has been a great challenge, especially in the energy sector, education, and health sector, but with the fact that this goal can be achieved with the smart grid and city concept, it is necessary to invest in these areas.

Depending on the smart city and network technological development, security problems should also be considered. Smart cities and networks have become widespread in recent years with developing technology, and risks and security gaps with developing technology are relevant. Smart grids cover electricity, water, and natural gas networks and their usage areas [1]. On the other hand, smart cities include smart transportation, smart stops, smart parking systems, and smart weather forecasting tools. covers all applications. All systems are managed from a single center. These central administrations are only possible with Supervisory Control and Data Acquisition (SCADA) systems. SCADA systems are defined as monitoring, controlling, and managing data from a single point. The data is sent to a single point, and at this point, meaningful information is obtained by making classifications on the data [1]. Data classification algorithms used in machine learning and artificial intelligence applications are preferred in the classification of data. In recent years, machine learning-based applications have been widely used in smart cities. The fact that the systems are smart and self-managed also offers comfort and convenience to people in city life.

Cyber-attacks offer an opportunity to deny availability and anonymity. In addition, it is difficult to determine who and who financed these attacks and which countries were behind these attacks. For this reason, it is challenging to detect risks and threats in cyberspace and take precautions against them. In such an environment, it is no longer mentioned providing absolute cybersecurity. Instead, it is aimed to make cybersecurity risks manageable and acceptable. It is recognized that being in an open and connected environment such as the Internet carries some risks associated with increased accessibility. It is imperative to be prepared for cyber incidents by managing these risks with a holistic approach involving all stakeholders and to ensure their continuity by eliminating these incidents with minor damage.

This chapter aims to assist studies that will carry out research, development, and design, especially on smart cities. First of all, the definitions of the concepts of smart city and machine learning are made. Machine learning based applications used in smarter cities are described. In addition, in this study, information about why smart cities are needed and their advantages-disadvantages are given.

10.2 Smart Cities

Smart cities are based on restructuring the situations of cities in a way that provides maximum efficiency with a focus on people and nature. In addition, smart cities have a human-oriented, strategic, development, change environment that creates and supports a management approach. Due to these reasons, these cities are city structures with improved service areas and living standards. These structures are based on creating new living spaces that are comfortable, healthy, people-oriented, self-sufficient, where resources are consumed efficiently and intelligently, respecting nature, minimizing environmental problems by using innovative and sustainable methods [2].

The smart city concept includes energy infrastructure, traffic management, waste management, healthcare, transportation, water supply, and other services. Thus, there is an interaction between service providers and citizens in a smart city. In a smart city, ICT is used to improve urban services' quality, productivity, and consistency. In addition, smart cities aim to reduce costs and resource consumption and improve communication between citizens and the government. Research on smart cities started in the 2000s, and many definitions were made for smart cities [3]. For example, a smart city can be defined as a combination of reliable infrastructure, data transfer quality, and corporate infrastructure. Nijkamp et al. noted that a smart city could be created under certain conditions. Classical and modern communication infrastructures should provide sustainable economic support and a high standard of living in terms of business, together with human and social capital in general. In addition, it is necessary to manage natural resources wisely through leadership [4].

The primary purpose of developing and popularizing smart cities is:

- Suitability of urban vehicles,
- The elegance of the city administration,
- Habitability of the living space,
- The intelligence of infrastructures,
- Long-term network security effectiveness [4].

Smart grids have gained significant importance as spreading in many countries and cities with technologies in recent years. However, these emerging technologies have weaknesses in terms of security. Smart grids include critical infrastructures such as transportation, education, finance, energy, and defense. However, for these structures to be part of smart cities, monitoring and managing them is of great importance [5].

This chapter address the security vulnerabilities that may occur on them. It has become one of the most critical steps to take security measures in smart grids. In this study, cybersecurity vulnerabilities in smart grids and measures to be taken against them are elaborated. These measures demonstrate the minimum requirements that each organization should undertake. It is noted that each organization may have significant losses due to cyber-attacks in terms of cost and data (Fig. 10.1).

In cyber-attack methods, the development of technologies, new hardware, and applications, and increased security deficits are observed. There are differences in the methods of attack against the weaknesses that occur with this increase. Especially as the method of cyber-attack, most malware, Distributed Denial-of-Service (DDoS), and advanced persistent threats show an increase in recent years. In this chapter, it is attempted to search for frequently used cyber-attack methods encountered during this research.

In all systems used in smart grids, the data being continuously operational, accessible, confidentiality, and integrity are the basic building blocks of information security. In addition, if smart networks can be monitored adequately and can adapt to the technological infrastructure as well as have expert staff for managing these systems,

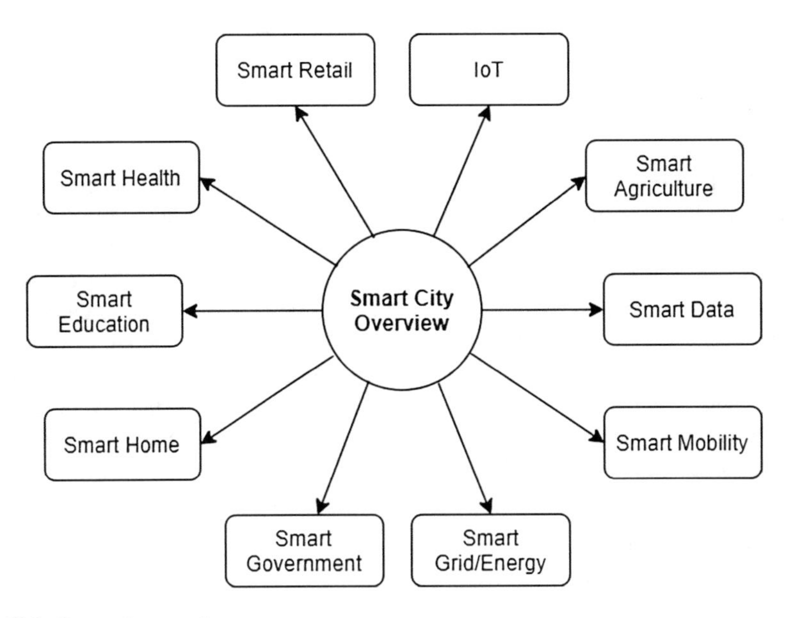

Fig. 10.1 Smart city overview

the attacks can be overcome. Human, technology, and company are the basic building blocks of success in information security. These basic building blocks are crucial for the successful management of smart grids.

10.2.1 Smart City Technologies

Smart cities generally use IoT devices as a combination of software application solutions, interfaces developed for users, and communication networks for the secure transmission of data. In the simplest definition, the IoT network is defined as a network of interconnected devices that can communicate and exchange data among themselves, such as vehicles, smart devices, sensors, or smart home appliances. The data in these networks is collected by sensors and devices and transmitted. In addition, this data is stored on servers or in the cloud and is used quickly when needed. The easy connectivity of these devices to the network and easy data analysis in terms of data makes it easy to bring together the physical and digital elements of the smart city. Thus, it increases the efficiency of both the public and private sectors, provides economic benefits, and improves the living conditions of society. Edge Computing enables the transmission of relevant information sent on the network by transferring data and communicating with each other throughout the communication.

10.2.2 Features of Smart Cities

Emerging technological trends such as IoT, automation, and machine learning enable people to use and adopt smart cities. Theoretically, it can be technologically included in the smart city system for any area of the management of cities in every aspect. For example, drivers without traveling for long periods in cities that are large in capacity and vitally crowded. In addition to smart parking meters that will help them find suitable parking spaces for their vehicles, smart meters that will allow digital payment in terms of shopping can be easily applied in smart cities. Moreover, in the field of smart transportation, smart traffic management is used to monitor traffic lights from a single center and analyze the data, to optimize the lighting lights on the streets, and to prevent too much congestion on the roads used by society according to certain hours of the day or rush hour schedules.

Smart public transportation systems are known as another technological aspect of smart cities. Intelligent public transport companies can coordinate the services they provide to the community and meet the needs of drivers in real-time to continuously increase efficiency, sustainability, and driver satisfaction. In addition, car-sharing, a trend in recent years, is rapidly increasing as a standard service in a smart city [6].

Energy savings and efficiency in smart systems have become the focus of smart cities. Smart sensors and smart streetlights are used in smart cities; When there are no cars or pedestrians on the primary vehicle and pedestrian roads, it can contribute to savings by getting dark. Smart grid technology also contributes to improving maintenance and scheduling for systems, saving on-demand power, and monitoring power outages.

Applications used in the smart city; aim to monitor, report, and take precautions against environmental risks such as waste and hygiene management, smart water technology, climate change, and air pollution in cities. The use of smart city technology is becoming more widespread every day to the extent that technological possibilities allow order to increase public safety, from monitoring high crime areas at specific points in city life to rapid emergency preparation with the help of sensors. For example, smart sensors; the early warning system before flood, landslide, hurricane, or drought, helps take the necessary precautions on time [6].

When evaluated in terms of smart buildings, this is generally a part of the smart city project, and the increase in such buildings in smart cities will increase security and living comfort. Intelligent sensors detect the wear and tear in the buildings making it possible to notify the authorities directly when these problems need repair [7]. Sensors used in smart cities can also detect existing leaks in water networks and other pipe systems used for different purposes from city life. In addition, these determinations help the firm reduce its economic costs and increase the efficiency of public services. In addition, smart city technologies; In addition, it enables job creation, space management, energy efficiency, and the use of fresher products for consumers. In addition, it contributes to urban production and urban farming in terms of efficiency [8].

As the population growth in smart cities continues to increase, these urban areas need to adapt to the increasing population by using their existing infrastructure and existing assets more efficiently [9]. In this respect, smart city applications can be improved depending on technology. It can advance smart city operations in terms of sustainability and improve the quality of life among community residents. In addition, smart city applications enable the cities of countries to find value again and create value for existing infrastructure services with the development of technologies. These improvements will help governments and citizens save energy, streamline new revenue models streams, and operational efficiency as a process.

Systems used by the municipality as they come into smart systems are expected to reduce the number of vehicles. Autonomous vehicles or self-driving cars, not about the necessity of having a car from a user's perspective, can potentially change the perspective of the population. The use of autonomous vehicles in terms of adoption society would lead to a significant reduction in the number of vehicles owned by, and thus to decrease the number of vehicles on the street is expected to be further reduced by the harmful gas emissions from the vehicle [10].

10.2.3 Why Do We Need Smart Cities?

The overall primary purpose of the smart city is to sustain economic growth and improve the high quality of life of the community. In addition, smart cities are to create a sustainable and green urban environment. It fills an essential advantage of smart cities, the citizens of their ability to deliver more cost and less infrastructure. As the population of cities continues to increase, these urban areas need to use their infrastructure and assets more efficiently to accommodate more people. Smart city applications make it possible to make such developments and increase city activities and citizens' quality of life. Smart city applications also make it possible to discover new values and create them using the existing infrastructure. In this way, new income methods can be created, and operational efficiency can be increased, allowing both governments and citizens to save more money [11].

10.2.4 Smart City Examples

Smart city ecosystem: Devices (Edge), the visible face of a smart city, consists of 3 technological levels: a core and a communication channel. In the edge layer; There are devices like sensors, cameras, other IoT devices, and smartphones. Baseline, which processes data from Edge and is a technology platform that analyzes. The link also provides two-way communication between the core and stable Edge to integrate seamlessly with various ecosystem components [12].

Smart cities use IoT devices as a combination of software application solutions, UI for users, and communication networks to communicate data. The IoT is a network that communicates with each other of various devices such as cars, sensors, or home appliances used in smart homes that can exchange data among themselves.

IoT data is collected by sensors and devices and transmitting data stored in the cloud or server. Data analysis and communication of these devices facilitate the interconnection of physical and digital elements of the city, thus maximizing the efficiency of both the public and private sector, improving citizens' lives, and providing economic benefits. IoT devices are also "Edge Computing" name that has processing capability. The paradigm of this calculation provides only the most critical and timely information transmitted over the network (Table 10.1).

The smart city concept has started to be implemented in many countries of the World. Examples of these are the cities that use the concept of smart cities, some of which are the most developed in terms of development. These cities are:

It is seen that most of the projects on the smart city are applied intensively in the Middle East countries and the Chinese state. However, in 2018 research, Reykjavik and Toronto were listed as the World's top four smart cities, along with Tokyo and Singapore. In Singapore, IoT-enabled cameras are preferred to detect the density of crowded environments while cleaning public and common areas. It often uses sensors and IoT-enabled cameras to monitor the movement of registered vehicles

Table 10.1 Smart cities examples in world [13]

Countries	Cities
Australia	Melbourne
United Arab Emirates	Dubai
China	Hong Kong
Missouri	Kansas City
California	San Diego
Ohio	Columbus
USA	New York
Canada	Toronto
Iceland	Reykjavik
England	London
Austria	Vienna
Spain	Barcelona
Japan	Tokyo

that are also used locally in the city. The development of smart technologies, the ability to track the energy use of companies and homes, helps to monitor and control waste management and water use in real-time. To give an example from city life in the city of Singapore, full-size robotic buses and aged monitoring autonomous vehicles are being tested to ensure the health and well-being of the elderly.

Kansas City's smart cities initiative, smart streetlights along the two tram routes in the city, interactive kiosks, and contains more than 50 free public Wi-Fi internet services. Available parking spaces in the city, traffic flow, and pedestrian points are open to anyone to use the data visualization application.

In San Diego, 3200 smart sensors were installed in early 2017 to optimize transportation traffic and parking, ensure public safety, improve general residents' living conditions, and raise environmental awareness. This city, to provide power to the electric vehicle charging stations with solar energy use has on employees and the cameras connected to this system helps to detect for detecting and monitoring the traffic offenders.

The United Arab Emirates, the Dubai-car navigation systems, town and city parks are used for training. Smart systems in the city, as well as health, education, and smart, intelligent readers also tourism.

Barcelona has also intelligent transportation systems and travel used smart bus systems as well as air temperature, pollution, and using noise and air and humidity sensors to monitor rainfall levels.

With the spread of smart cities, while living standards will improve, especially in crowded places, the problem of time in city life will also disappear. With more qualified and livable cities, many confusions will be prevented, and a lot of energy and time will be saved.

10.3 Machine Learning Applications in Smart Cities

While artificial intelligence aims to enable computers to mimic human behavior and cognitive processes, machine learning is the branch that focuses specifically on developing algorithms and applications that learn from data and automatically improve accuracy over time through experience. Machine learning can have a massive impact on smart cities and IoT networks, where many connected devices collect and share inhomogeneous data. This is a treasure trove for improving existing urban services and developing new smart applications.

10.3.1 Smart Waste

Municipal solid waste collection is economically inefficient and rarely effective. Artificial intelligence and machine learning algorithms are integrated to improve the performance of the solution further. The first step is to equip the tanks with sensors to monitor data on fill level, date and time of last waste collection, and exact location. This allows operators to only ship trucks when and where the holds are close to full, reducing the number of truck rolls and their associated mileage, which reduces pollution and congestion. The compartment fill level prediction system utilizes a deep neural network trained on a dataset containing historical data from smart compartments [14].

10.3.2 Continuous Learning

More complex logic will soon be integrated to teach the bin to identify nearby points of interest such as supermarkets, stadiums, stations or hospitals, and to predict waste input based on variables such as density of people, events, festivals, and more. Continuous learning techniques allow the model to evolve by incorporating new data and accumulated knowledge continually. This means that operators can make data-driven decisions about container quantity, capacity, and location by analyzing filling patterns. Trash bins can be positioned to suit different urban scenarios and adapted within a few days to changing conditions, i.e., habits or behaviors of different residents, such as the increase in packaging-related garbage due to increased e-commerce and home deliveries during the pandemic. With the Smart Urban Network and machine learning algorithms, cities can truly improve waste collection and management, increasing livability while protecting public health and the environment [15].

10.3.3 Asphalt Condition Monitoring

The basic raw material used in the pavement of highways today is asphalt. However, due to its structure, asphalt wears out over time depending on usage and environmental conditions, and situations such as deterioration, collapse, and cracking occur on the surface. This is very important to detect and repair these wears without causing harmful consequences, especially in driving safety [16].

10.3.4 Air Pollution Forecast

One of the most critical issues in smart cities is the determination of the air quality of the city and the measures to be taken against it. Significant work has been done on this subject with machine learning. The current leading practices in the field of health are given below [17]:

- Collection of health data (electronic health record),
- Telemedicine applications (telemedicine),
- Mobile health applications (mhealth),
- Location tracking,
- Elderly care of the elderly,
- Monitoring of chronic patients (stroke, diabetes, Alzheimer's) (chronic patient monitoring),
- Preventive healthcare practices (Preventive healthcare).

10.3.5 Vehicle Routing in Smart Cities

The vehicle routing problem discussed is examined under the assumption that vehicles can get traffic and average speed information from other vehicles in the city before they move from the starting point [18].

10.3.6 Effective Transportation Planning

Effective transportation planning in smart cities has become essential in developed societies. Effective planning of transportation in crowded societies of smart cities offers significant added value in terms of comfort and economy in daily life [19].

10.3.7 Forecasting of Air Quality

Smart cities are using measurements of air quality necessary measures can be taken. In this study, air quality is measured by machine learning. The results of the air quality measurement are communicated to the relevant institutions, and the actions to be taken are determined by preparing in advance [20].

10.3.8 License Plate Recognition System in Smart Transportation

In recognition of the smart grid plates on vehicles used, machine learning and neural networks have become possible. These applications are used in vehicles with license plate identification systems at the entrance to shopping malls, parking lots, and institutions [21].

10.3.9 Tourism Developing

Smart tourism applications are being developed in smart cities thanks to machine learning algorithms. These developments provide significant contributions as points of choice in tourism in smart cities [22].

10.3.10 Intelligent Parking System

In smart cities, smart parking systems are critical in crowded communities to determine how many parking points are in which parking lot. Thanks to its smart parking points, it offers comfort and regular life in cities [23].

10.3.11 Density Monitoring Systems

Density monitoring systems in smart grids prevent complexity in the city. In addition, thanks to these monitoring systems, it is possible to offer a safe and comfortable life to society [24].

10.3.12 Smart Stall Systems

In today's world, where time is valuable, project studies have accelerated passenger satisfaction. Kayseri Metropolitan Municipality, which uses the latest innovations in technology more and more in its services, provides modern, fast, high quality, and accurate transportation with this system [25].

10.3.13 Barrier-Free Smart Transportation

Considering the living standards it offers, the population of big cities is increasing day by day. In direct proportion to the city's population, traffic problems arise, and the time spent on transportation increases. In addition to all these negativities, the problem becomes inextricable when it is considered that the individual who wants to provide transportation has a physical disability [26].

10.3.14 Smart Real Estate Management in Smart Cities

Smart real estate management is vital in comfort and facilitating life in smart cities where society is densely populated. To meet human needs, financial learning and artificial intelligence applications are developed in this area and make significant contributions to society [27].

10.3.15 Smart Hotel

The smart hotel is a labor-intensive industry, but with the support of artificial intelligence technologies, the services provided by the employees to the guests are faster, more effective, and provide benefits that will increase guest satisfaction [28, 29].

10.4 Discussion

Most of the materials listed above, consult with each institution, pay attention, and are among the most critical issues. Organizations do not want results that will cause material and reputation loss by experiencing cybersecurity incidents. These points can be expanded, but in this study, they are grouped based on central themes. As technology develops, smart grids and cities may become more resistant to cyberattacks, but care must be taken to be local and national.

Data thieves can easily exploit vulnerabilities in smart systems such as security systems and facial recognition cameras. They can also be easily compromised and commit serious crimes that threaten society. In such studies, we say that security vulnerabilities will be more attractive to hackers than a special system. However, since all systems in smart cities are interconnected, hacking any system can cause many domino messes. For this reason, systems should be considered as a whole in terms of security strategy. In addition, all security vulnerabilities should be minimized for each used system and these vulnerabilities should be eliminated [24]. A security system known as a firewall is also required to protect, monitor, and control network traffic used by computer systems. Firewalls prevent unauthorized access to data. Network or cities using IoT technologies, smart cities in the network ensure the continuous transmission of data.

With the increase in big data and wireless communication in smart cities, cyber-attacks carry a high risk in terms of data security. It should aim to create a full-fledged smart city while using safe and smart traffic systems on the main streets, spreading it to all smart cities and making the life of the city residents easier. However, you should consider the chaotic situations it can cause when a data thief attacks or hijacks traffic monitoring sensors or the city's running system. In terms of smart city management, cities have plans for natural disasters such as floods and earthquakes, especially due to research by security research companies. However, many city governments should be warned about planning for cyber-attacks. It is necessary to think that a human-centered disaster can cause great destruction and develop smart strategies against cyber-attacks.

The use of IoT systems in these areas and the communication of everything with each other, the development of big data and artificial intelligence systems, and cybersecurity risks increase at that rate. Despite these measures, institutions and management units should cooperate and implement these recommendations. It is complicated to talk about 100% security in countries unless countries produce tools, equipment, and software locally. For this reason, countries should invest in this direction, and states should lead researchers and institutions on these issues.

10.5 Conclusion

Smart grids and cities have become widespread in many countries in recent years. With widespread use, the communication and connection methods of IoT systems have gained importance. Companies that produce these technologies have to follow the constantly developing technologies. They should produce devices according to the smart system integrations to be made for each system produced. They should also prioritize the security of the systems. These systems must have international certificates.

Machine learning-based applications used in smart cities are applications that use smart stops, smart waste, air pollution, and quality forecasts, smart traffic lights, smart streetlights, smart parking lots, smart energy management, smart healthy and smart hotels. These applications are integrated into IoT-based systems, resulting in the emergence of comprehensive data. This data is called big data and it is one of the most critical issues to protect, manage and produce meaningful data by processing this data. In addition, it is important to protect personal data according to General Data Protection Regulation (GDPR) laws and to take certain security measures against attacks. The most important features of smart grids are that they can be monitored, controlled, data processed, and managed sustainably. For this reason, it is not enough to use IoT-based systems for cities to be smart because, at the same time, backup systems are needed for these systems to work continuously in case of a breakdown or problem.

The combination of machine learning and IoT systems raises security problems. Cyber security measures should be taken at the highest level to protect the systems. In terms of data protection, these measures should be considered on a systemic basis. Studies on this subject should be evaluated in terms of security risks and more studies should be included on how to protect data.

References

1. Özarpa, C., Aydın, M.A., Avcı, İ.: International security standards for critical oil, gas, and electricity infrastructures in smart cities: a survey study. In: Ahmed, M.B., et al., (eds.) Innovations in Smart Cities Applications, The Proceedings of 5th International Conference on Smart City Applications. Lecturer Notes In Systems And Network, vol. 4, chapter 89. Springer, Berlin (2021)
2. Avcı, İ: Investigation of cyber-attack methods and measures in smart grids. J. Sci. **25**(4), 1049–1060 (2021)
3. Chourabi, H., et al.: Understanding smart cities: an integrative framework. In: 2012 45th Hawaii International Conference on System Sciences, pp. 2289–2297. IEEE (2012)
4. Nam, T., Pardo, T.A.: Conceptualizing smart city with dimensions of technology, people, and institutions. In: Proceedings of the 12th Annual International Digital Government Research Conference: Digital Government Innovation in Challenging Times, pp. 282–291 (2011)
5. Avcı, İ., Özarpa, C., Aydın, M.A.: Mitigating global warming in smart energy grids via energy supply security for critical energy infrastructure. Int. J. Glob Warming **25**(3/4) (2021)
6. Badii, C., Bellini, P., Difino, A., Nesi, P.: Smart city IoT platform respecting GDPR privacy and security aspects. IEEE Access **8**, 23601–23623 (2020)
7. Hanes, D., Salgueiro, G., Grossetete, P., Barton, R., Henry, J.: IoT Fundamentals: Networking Technologies, Protocols, and Use Cases for the Internet of Things. Cisco Press, Indiana (2017)
8. Ullah, Z., Al-Tudjman, F., Mostarda, L., Gagliardi, R.: Applications of artificial intelligence and machine learning in smart cities. Comput. Commun. **154**, 313–323 (2020)
9. Nambiar, R., Shroff, R., Handy, S.: Smart cities: challenges and opportunities. In: 2018 10th International Conference on Communication Systems and Networks (COMSNETS), pp. 243–250. IEEE (2018)
10. Yaqoob, I., Khan, L.U., Kazmi, S.A., Imran, M., Guizani, N., Hong, C.S.: Autonomous driving cars in smart cities: recent advances, requirements, and challenges. IEEE Netw. **34**(1), 174–181 (2019)
11. Mohammadi, M., Al-Fuqaha, A.: Enabling cognitive smart cities using big data and machine learning: approaches and challenges. IEEE Commun. Mag. **56**(2), 94–101 (2018)
12. Jararweh, Y., Otoum, S., Al Ridhawi, I.: Trustworthy and sustainable smart city services at the edge. Sustain. Cities Soc. **62**, 102394 (2020)

13. Shafique, K., et al.: IoT for next-generation smart systems: review of current challenges, future trends and prospects. IEEE Access **8**(99), 1–1 (2020)
14. Alqahtani, F., Al-Makhadmeh, Z., Tolba, A., Said, W.: Internet of things-based urban waste management system for smart cities using a Cuckoo Search Algorithm. Cluster Comput. **23**, 1769–1780 (2020)
15. Kolomvatsos, K., Anagnostopoulos, C.: Reinforcement learning for predictive analytics in smart cities. Inf. Multi. Digit. PublishingInst. **4**(3), 16 (2017)
16. Baygın, M., Yaman, O., Tuncer, T.: Akıllı şehirler için özellik çıkarımı ve makine öğrenmesi tabanlı asfalt durum izleme yaklaşımı. Avrupa Bilim ve Teknoloji Dergisi **23**, 81–88 (2021)
17. Gültepe, Y.: Makine öğrenmesi algoritmaları ile hava kirliliği tahmini üzerine karşılaştırmalı bir değerlendirme. Avrupa Bilim ve Teknoloji Dergisi **16**, 8–15 (2019). https://doi.org/10.31590/ejosat.530347
18. Çimen, M., Soysal, M., Sel, Ç.: Akıllı şehirlerde araç rotalama: benzetimsel dinamik programlama tabanlı bir sezgisel yöntem, pp. 71–84 (2017). http://indexive.com/uploads/papers/pap_indexive15955052582147483647.pdf
19. Başkaya, O., Ağaçsapan, B., Çabuk, A.: Akıllı şehirler kapsamında yapay zekâ teknikleri kullanarak etkin ulaşım planlarının oluşturulması üzerine bir model önerisi. GSI j. Serie C: Advancements Info Sci. Technol. **3**(1), 1–21 (2020)
20. Dokuz, Y., Bozdağ, A., Gökçek, B.: Hava kalitesi parametrelerinin tahmini ve mekansal dağılımı için makine öğrenmesi yöntemlerinin kullanılması. Niğde ömer halisdemir üniversitesi mühendislik bilimleri dergisi **9**(1), 37–47 (2020)
21. Javadov, I.: Evrişimli sinir ağları ile plaka tanımada algoritmaların karşılaştırılması (doctoral dissertation) (2020)
22. Yalçınkaya, P., Atay, L., Karakaş, E.: Akıllı turizm uygulamaları. Gastroia: J. Gastronomy Travel Res. **2**(2), 85–103 (2018)
23. Uysal, E., Elewi, A., Avaroğlu, E.: Nesnelerin interneti tabanlı akıllı park sistemleri incelemesi. Euro J. Sci. Technol. **20**, 360–366 (2020)
24. Kızrak, M.A., Bolat, B.: Derin öğrenme ile kalabalık analizi üzerine detaylı bir araştırma. Bilişim Teknolojileri Dergisi **11**(3), 263–286 (2018)
25. Taşyürek, M., Çelik, M.: Akıllı durak sistemindeki araç seyahat sürelerinin birleşik yapay sinir ağları kullanarak tahmini. Avrupa Bilim ve Teknoloji Dergisi 72–79 (2020)
26. Hakverdi, F.: Akıllı şehirlerde engelsiz akıllı ulaşım (Master's Thesis, Necmettin Erbakan Üniversitesi, Fen Bilimleri Enstitüsü) (2020)
27. Önder, H.: Gayrimenkul 4.0 ve emlak yönetiminde dijitalizasyon. Elektron. Sosyal Bilimler Dergisi **20**(79), 1341–1357 (2021)
28. Tripathy, A.K., Tripathy, P.K., Ray, N.K., Mohanty, S.P.: iTour: the future of smart tourism: an IoT framework for the independent mobility of tourists in smart cities. IEEE Consum. Electron. Mag. **7**(3), 32–37 (2018)
29. Demir, Ç.: Konaklama işletmelerinin iş süreçlerinde yapay zekâ teknolojileri ve akıllı otel. J. Tourism Gastronomy Stud **9**(1), 203–219 (2021)

Chapter 11
Recent Developments of Deep Learning in Future Smart Cities: A Review

Nur Akmaliza Zanury, Muhammad Akmal Remli, Hasyiya Karimah Adli, and Khairul Nizar Syazwan W. S. Wong

Abstract Urban and rural areas are adapted to address the latest technological innovations such as smart cities and concurrently arisen new opportunities for public safety to citizens and visitors. Smart cities aim to sustain the emerging urbanization, amount of energy used, conserve green living, and simultaneously help enhance people's productivity and lifestyle. Apart from that, smart cities are the catalyst in improving people's ability to use advanced computerized information efficiently. Deep learning (DL) and the Internet of Things (IoT) are crucial in establishing government and industrial business standards, risk management, application, and productivity output. In smart cities, several IoT sensors have been installed in a some places for public data gathering such as road traffic and people's movement. DL is an advanced machine learning technique meant for extensively on data collection, patterns understanding, and data prediction. Besides, DL has caught the interest of every researcher around the world, and it has delivered remarkable results in comparison to the current standard techniques. This chapter aims to explore the recent development of DL techniques for the evolution of smart cities and to study the future directions.

Keywords Deep learning · Artificial intelligence · Machine learning · Smart city

N. A. Zanury · M. A. Remli (✉) · H. K. Adli · K. N. S. W. S. Wong
Institute for Artificial Intelligence and Big Data, Universiti Malaysia Kelantan, City Campus, Pengkalan Chepa, 16100 Kota Bharu, Kelantan, Malaysia
e-mail: akmal@umk.edu.my

N. A. Zanury
e-mail: akmaliza.z@umk.edu.my

H. K. Adli
e-mail: hasyiya@umk.edu.my

K. N. S. W. S. Wong
e-mail: nizar.w@umk.edu.my

M. A. Remli · H. K. Adli · K. N. S. W. S. Wong
Department of Data Science, Universiti Malaysia Kelantan, City Campus, Pengkalan Chepa, 16100 Kota Bharu, Kelantan, Malaysia

© The Author(s), under exclusive license to Springer Nature Switzerland AG 2022

G. Marques et al. (eds.), *Machine Learning for Smart Environments/Cities*, Intelligent Systems Reference Library 121,
https://doi.org/10.1007/978-3-030-97516-6_11

11.1 Introduction

Mobile devices and sensor-based apps have gotten more intelligent in recent years, allowing devices to connect. The number of connected devices overtook the inhabitants of the world in 2008 [1, 2]. All intelligent devices such as smartphones and sensors, are connected with one another, leading to the age of intelligent city. In the meantime, as the number of electronic devices grows, so does the amount of data they collect. Artificial Intelligence (AI) applications using Deep Learning (DL) techniques and Internet of Things (IoT) devices arise as new apps evaluate obtained data to make meaningful correlations and possible conclusions.

DL is a discipline of machine learning which also being called neural networks. It is well adapted to dealing with large amounts of data and complex computing jobs such as image processing, speech recognition, patterns understanding and data prediction [3]. In comparison to existing standard procedures, DL has caught the interest of every researcher in the world, and it has delivered outstanding results in processing vast amounts of data without being complicated or sluggish.

The neural networks consist of immense hidden layers, which are usually known as "deep" as visualized in Fig. 11.1. The quantity of hidden layers in deep learning neural networks is dependent on the total of calculated features. DL has the ability to estimate features automatically. Thus, before using the method, there is no need to analyze and extract the features [3].

From Fig. 11.1, each layer makes up the backbone of the deep learning models. Each layer has been assigned a specific processing job, and input may pass through the hidden layers several times to improve and enhance the final output [3]. The hierarchy of hidden layers in deep learning neural networks will undergo data processing and transforming by leveraging each of the output from the subsequent layer. Thus, producing better and more accurate outputs. Deep learning models enhance their capacity to make correlations and connections as more data is analysed, essentially learning from prior results.

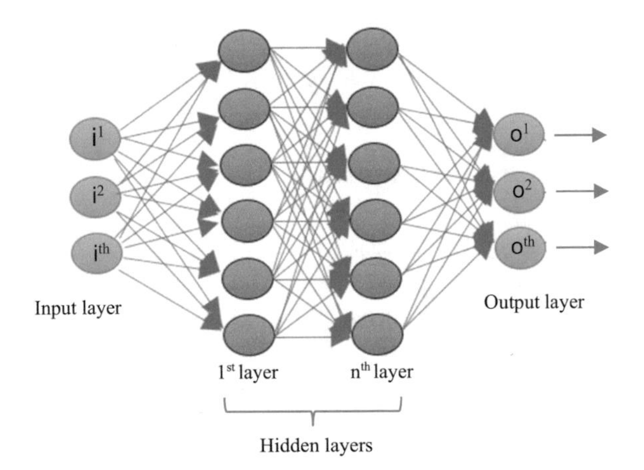

Fig. 11.1 Illustration of deep learning neural networks

IoT is an advanced modern technology that allows data to be transmitted across a network. According to a statistic, IoT is a thriving innovation that comprises interconnected devices, regardless of which devices as long as it is smart applications. The integration of various devices within a connected network facilitates IoT. Nevertheless, in 2011, the number of connected devices was reported to increase, despite the number of inhabitants being lower [4]. It comes from a variety of devices, apps, detectors, and internetworking that allows data transfer and transmission [1, 5].

In the idea of IoT, smart applications are said to be linked with one another using some analytical tools with respect to their geographical area [6]. Service domain applications have already been using IoT devices on a daily basis, for example, to control air pollution, noise pollution, automobiles movement, including monitoring systems. In addition, IoT enhances and trigger the different aspects of individual lifestyle by providing cost-effective community services, promoting better public transportation, eliminating heavy traffic, and saving one's life [6].

Figure 11.2 exhibits the main focus of IoT applications within smart cities. Using IoT platforms in homes and buildings, different equipment allows for automating identical and routine tasks. Indeed, by transforming objects into data substantially connected by using the Internet, services may be delivered using web interfaces. The data transmission between smart applications and the Internet enable the smart applications to be controlled remotely [6, 7]. Smart lighting technology is one of the intelligent innovations already used globally, especially in urban areas [8, 9]. This technology enables people to control the lighting even from a distance. Furthermore, it is reported that at least 19% of energy consumption is utilised and wasted without smart lighting technologies, resulting in air pollution increase [10]. However, smart lighting technology can save roughly 45% of the energy required for lighting [9]. Meanwhile, from other concepts, sensors can be deployed in various cities to collect and analyze public data for enrichment [11].

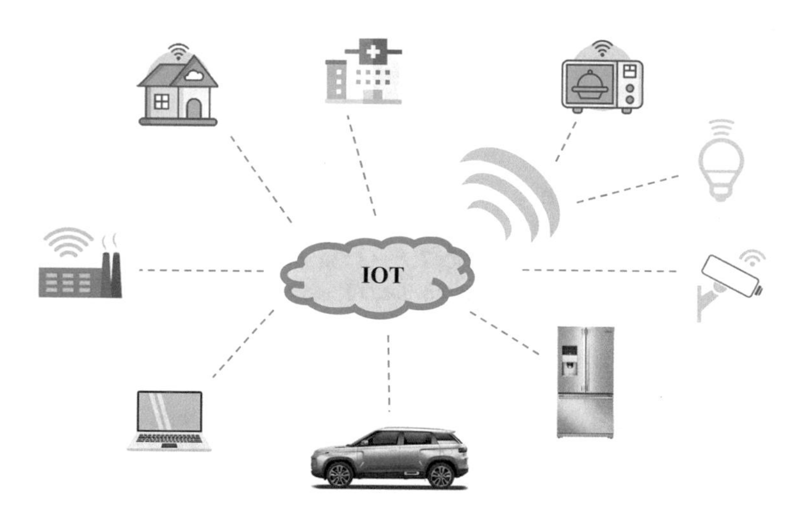

Fig. 11.2 Illustration of IoT applications

Smart cities are among the newest advanced technology that urban and rural areas have adapted to solve. Smart cities strive to effectively manage urbanization, energy consumption, environmental sustainability, economic and living standards of people. Specifically, the concept of "smart cities" promises to tackle several urgent issues that come with rapid urbanization changes, such as less population in urban areas, heavy traffic, energy consumption system, and poor drainage system [12]. Furthermore, it is unsurprising that IBM, Cisco, Telekom, Toshiba, and Google are just some of many innovative organizations actively investing in the evolution of smarter cities [12].

Municipalities and administrative decision-makers, and non-profit and civic organisations should be aware of the advantages of smart cities. The energy revolution's social and political necessities and the exciting possibilities of an interconnected world within the IoT environment are hastening the transformation of cities into smarter cities. Furthermore, smart cities hold enormous promise for practically all businesses in the fields of information technology, real estate, telecommunications, energy, transportation, detector systems, and analytical data. As a result, many startup companies are concurrently entering advanced technology industries like the IoT and artificial intelligence, placing competitive pressure on established firms. The smart city concept has become a reality for IoT and artificial intelligence innovations [13–15].

The goal of smart cities is to provide adequate service to people by utilising modern technologies and analytical data on data acquired by IoT [6]. This can be leveraged by deploying several IoT sensors in various cities to gather public data such as road traffic, water systems, and people's mobility [15]. These actual and live data gathered can be utilised to manage resources and assets more efficiently in the future. In addition, the intelligence of a city is influenced by the technology-driven infrastructure, green living challenges, intelligent public automobiles, and many other advanced technologies that are implemented around [14, 15]. The smart city concept is also driven by the growing number of inhabitants across the world and rising urbanism, which is expected to increase at least 10% in the next three decades [16, 17], leading to 70% living in urban areas by 2050 [12, 17]. Countries all over the world are striving to equip their urban areas to cope with large-scale population inflows [16, 18].

By 2040, urban areas will be home to 65% of the global inhabitants. At least two people are estimated to move into urban areas every second from all around the world. Meanwhile, in the coming 2050, 70% of the global inhabitants will move into urban areas [12, 17]. As a result, urban areas have a higher population than rural ones.

This chapter comprises four sections. The first section presented a comprehensive review for future smart cities concepts and the advanced technologies approach to urban sustainability that this concept implies. This section is followed by a further study of the recent development of DL techniques in smart cities (Sect. 2). These further reviews are focused on the numerous domains in smart cities, and the recent development of DL techniques in future smart cities. Section 3 focus on the future direction of DL in smart cities. Finally, this chapter concludes with a critical analysis of the current smart cities' development theme, which plays a critical role in bringing consciousness to the research gaps. In summary, this chapter aims to survey the recent development of DL techniques in the future evolution of smart cities.

11.2 Development of Deep Learning Research in Smart Cities

The majority of the researchers have been applying DL techniques in numerous domains, such as transportation, services, nature, and the protection of the general public. This chapter reviews the recent development of DL techniques with the integration of diverse artificial intelligence applications in future smart cities. These applications are generally known as "smart" or "intelligent" as they utilize artificial intelligence technologies to obtain appropriate and useful information from advanced analytical data in order to better comprehend cities [19].

Figure 11.3 shows several smart city domains categorized into various modern technologies of artificial intelligence applications worldwide.

Traffic issues are a crucial research area in the urban areas and smart transportation. Traffic issues are a critical concern in any city due to the increasing number of populations using vehicles, leading to heavy traffic flow and accidents [21]. Besides, traffic issues can also cause air quality pollution and excessive fuel consumption. Nevertheless, some other researchers worked on parking detection systems and bus passengers' destinations prediction systems.

Many types of research have been made related to the Artificial Intelligence (AI) field, such as machine learning, deep learning, and advanced data analysis, solving

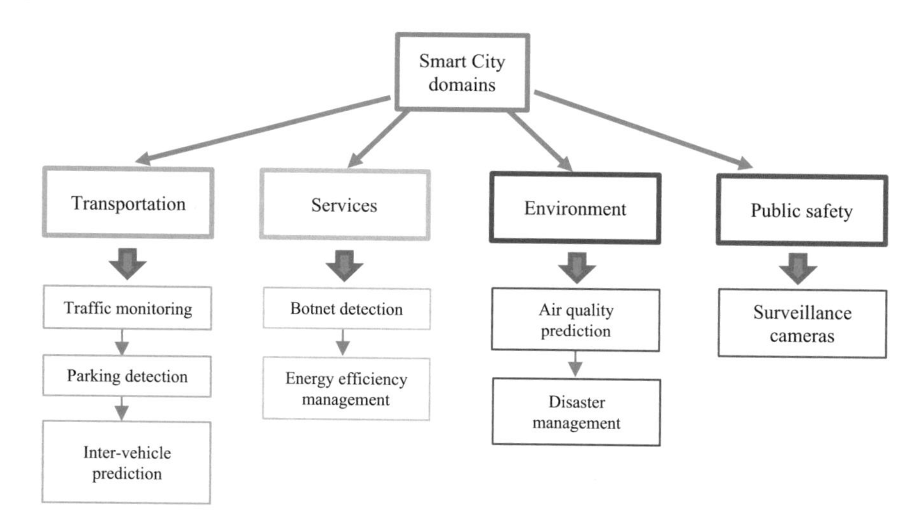

Fig. 11.3 Classification of the smart city domains

problems more manageable and producing more accurate forecasting systems. For instance, in [20], a comparison between random forest, AdaBoost, and logistic regression have been made to propose a traffic prediction system. These three approaches are analyzed and compared to find the most accurate machine learning algorithm. At the same time, the authors also proposed a traffic congestion reduction system by detecting heavy traffic flow and diverting incoming automobiles to the shortest or optional route that is accessible.

Another research presented in [21] proposed developing and implementing the same aspect of an artificial intelligence application in predicting the traffic flow within a city. Nevertheless, this research uses multiple linear regression (MLR) for pre-processing stage and artificial neural network (ANN) for predicting the traffic flow within a city. In [22], the authors proposed an inter-vehicle distance prediction system using a deep neural network with the integration of IoT. To compute average vehicle velocity and density over the next five minutes, they used a deep neural network to collect and predict huge amounts of road traffic data. The suggested framework's cars will be able to track their location via GPS. They will switch data with other vehicles around them about their current speed, motions, and also position whilst still uploading the data to the server. Additionally, accidents can also be avoided by alerting other vehicles to a rapid change in speed or sending heavy road traffic flow information to other vehicles to improve guidance services. For instance, each vehicle on the road may compute the average of the other vehicle density without using potential existing devices and simultaneously able to reduce the control overhead.

Besides that, [23] covered a smart parking detection system using a convolutional neural network (CNN) to classify the parking lot occupancy. CNN is a decentralised system for visual parking lot occupancy detection explicitly created for smart cameras utilised as a decentralised solution. These smart cameras were generally installed on the property closest to the parking area and placed in an external camera box. Smart cameras continuously take photos of a parking lot area and monitor the occupancy status of each parking place.

Nevertheless, the smart city can be considered a service organization with citizens as the customers, and the smart city itself will provide services to its citizens. Since DL models are employed with IoT devices, advanced and sophisticated assaults through an online network such as botnets are frequent. Apart from applying DL techniques to artificial intelligence applications, DL also can be applied in preventing existing artificial intelligence applications from networks attacks. In [24], the authors use DL for the classification and detection of the domain name into corresponding domain generation algorithms. The proposed research utilizes self-learning approaches that may be quickly developed and evaluated to identify advanced persistent botnet threats.

Other research presented by [25] employs deep neural networks and random forest to extract and develop forecasting models of certain energy usage across a city's public government organizations. The forecasting models are then implemented in artificial intelligent applications to collect and analyse every energy consumption. The authors also concluded that the proposed research had shown the capability of machine learning approaches in energy conservation within public government organizations.

Massive global urbanisation has brought significant challenges to our living environment—these challenges include air pollution, severe weather, and even earthquakes. The continual monitoring and preservation of the environment are some of the most critical elements of smart cities [19]. By undergoing advanced continuous analysis on public environment data resources, the smart environment promises to produce better monitoring and forecasting systems. With smart environment monitoring systems, smart cities can now predict and avoid any catastrophic disasters. This chapter will be focusing on air quality prediction and disaster management.

In [26], the authors proposed a DL model to address air pollution issues in smart cities. They configured the network with the best hyperparameters which are “input sequence”, “hidden layers”, “low short-term memory (LSTM) units”, and “the batch size” to build the most optimal LSTM structure. The proposed DL model is then generated and validated using mean absolute error, and root mean squared error metrics. The authors concluded that the overall findings confirm that their proposed LSTM-based model outperforms the competition. As a result, the findings indicate that applying an LSTM-based prediction model to IoT data is compelling and viable for a smarter city.

Apart from that, natural and man-made disasters are becoming extremely serious. Unfortunately, because of the stiff design of building construction and the high population in urban areas [16, 17], and high traffic flow [21], smart cities are disproportionately affected. Hence, the disaster management system in a smart city is a critical aspect to be taken care of. This disaster management system attempts to more adequately and proficiently address crises. [27] claim to be the first to use DL techniques in the disaster management system, with the help of smart traffic systems in predicting urban traffic behaviour. The disaster management system is implemented by forecasting the traffic flow within the cities [27].

In a catastrophe emergency, road traffic management is thought to be critical in escaping the impacted region and monitoring road traffic flow in all other areas of cities to avoid overcrowding and road blockages. Furthermore, GPS data can be employed to track and record the total number of vehicles going through a given location simultaneously and direct people to other routes to stay safe, avoid excessive traffic, and efficiently offer emergency services in the impacted area [27].

When a disaster strikes, smart public safety technologies play a vital role in recovery. Surveillance cameras mounted on neighbouring streetlights and private properties could deliver additional perspectives on the scenario [28]. Whilst smart cities have made substantial advancements, achieving real-time situational awareness and cross-agency engagement vital for citizen safety and security remains a long way off [28]. Smart cities must implement cutting-edge technologies that make it easier to prevent and respond to problems faster and more cohesive way to combat crime, counteract possible terrorist attacks, and create safer places for people.

In March 2018, Fujitsu launched its new smart city monitoring application known as Citywide Surveillance V2 [28]. Citywide Surveillance uses deep learning techniques in recognition and attribute classification based on video or photo images of individuals or vehicles. For example, vehicles can be recognized by their type, manufacturer, colour, and vehicle registration number. Also, individuals can be recognized by their types and skin colours, including clothing and facial characteristics. The installation of numerous surveillance cameras around the smart city and the produced images and video data resources could vastly improve several crucial issues, including traffic safety issues.

The smart city domain consists of multiple domains such as transportation, services, environment, public safety, healthcare, and education. As summarized in Table 11.1, this chapter focuses on smart transportation, smart services, smart environment, and public safety monitoring. Each of these domains then either proposes the same artificial intelligent applications or different artificial intelligent applications.

Smart transportation is the most common research done, followed by environmental measures and public services. Moreover, smart transportation focuses primarily on monitoring road traffic flow to avoid congestion and road blockages in one place. Besides, in crowded places, accidents are likely to occur; for example, a car from behind may collide with another vehicle in front of them. Therefore, with the help of deep learning technology to predict the distance between vehicles and each vehicle density through big traffic data, these kind of issues can be avoided.

In addition, accidents can be avoided by alerting other vehicles of rapid speed changes or sending heavy road traffic flow information to other cars to improve guidance services. Other than that, sometimes users may find it hard to find parking in crowded places. Thus, deep learning technology like smart cameras can assist users in monitoring the occupancy status of each parking IoT.

11.3 Future Direction of Deep Learning Research in Smart Cities

DL is gradually transforming how urban centres manage, assist, and sustain all potential amenities such as mobility, power, medical, and networking. Incorporating deep learning and machine learning technology into smart cities could take numerous different ways in the future. It is assumed that a training model generates dependable results when the training and testing data have comparable feature sets and distribution models. To increase the interaction between smart gadgets and consumers, researchers should incorporate semantic concepts into smart city applications.

This chapter has listed many research subjects that have been widely researched by smart cities researchers, relying on the detailed overview of smart cities before. On the other hand, smart healthcare is still a neglected and unexplored area in research. Healthcare is often one of the essential services within the cities, despite whether the industry's primary producers come from public companies or governments [29]. Even though the availability of remote medical care services is rising as technology progresses, this problem is scarcely spoken about in this research area. In South Korea, high-tech healthcare is an important research and development services sector

Table 11.1 Outline of DL models for several smart city domains

Domain	References	Model	Result
Smart transportation (traffic)	[20]	Random forest	Avoiding congestion level, by diverting the incoming automobiles to the shortest or optional route that is accessible. Results also show that the proposed research can identify the traffic level Forecasting the road traffic flow and suggesting optional routes have proven to eliminate the heavy road traffic flow and also the air pollution
Smart transportation (traffic)	[21]	Artificial neural network (ANN)	The architecture for traffic prediction based on neural networks and strategies such as machine learning, computer vision, and deep learning was made
Smart transportation (driving)	[22]	Temporal convolutional network (TCN)	Enable users to predict inter-vehicle distance and each vehicle density through the traffic big data without incurring high communication overhead
Smart transportation (parking)	[23]	Convolutional neural network (CNN)	In comparison to PKLot, CNRPark-EXT outperforms and deduce the highest performance techniques. The CNN architecture performs similarly to AlexNet in the parking space occupancy identification task
Smart service	[24]	Deep learning	The results reveal that deep learning-based botnet detection beats traditional machine learning-based approaches. The proposed approach can be used to monitor all Internet-connected IoT devices at the ISP stage

(continued)

Table 11.1 (continued)

Domain	References	Model	Result
Smart service	[25]	Deep neural network, random forest	The Random Forest model was shown to be the most precise on validation data, with a SMAPE of 13.5875%, indicating the effectiveness of machine learning approaches in energy management in a government organization
Smart environment (air quality prediction)	[26]	Recurrent neural network (RNN), Long Short-Term Memory (LSTM)	According to the findings, the application of an LSTM-based prediction model to IoT data is successful and promising
Smart environment (disaster management)	[27]	Convolutional neural network (CNN)	Demonstrate the efficacy of a DL technique in disaster management and accurate traffic behaviour prediction in emergency scenarios
Public safety (surveillance cameras)	[28]	Deep learning	Developed Citywide Surveillance uses recognition and attributes classification based on video or photo images of individuals or vehicles

in programs within developing cities. Furthermore, the recent COVID-19 outbreak demonstrates the importance of science and health research and smart technologies in coping with epidemics in urban areas [30]. As a result, more thorough research on healthcare services in urban areas should be emphasized in the eventuality.

Furthermore, one of the most essential areas addressed in smart city studies is air quality monitoring. Air quality monitoring is becoming one of the most critical challenges in developed countries like China and Europe because improved air quality can contribute to people living healthier and better lives [29, 30]. While air quality monitoring has been addressed in certain environmental sensing and control research, this has not been extended towards the degree of successful handling that other smart city research that focuses on "management" has. This could be owing to a scarcity of increased pollution information, which makes extensive academic research difficult. Hence, more research and development concerning air system quality should be overseen, especially in urban areas.

On the other hand, smart city research pays less attention to the issue of the public and society. Local governments and public businesses, for example, have figured prominently in the delivery of services in urban centres. Many public areas are key themes in smart city services, including energy usage, traffic, water quality, and trash management. Smart cities research does not assure the long-term sustainability of smart cities development.

Smart parking, road traffic monitoring, and electric car research, for example, could lead to more convenient auto-dependent communities that are generally regarded as less strategic. As a result, more research on public services and local communities is required in all study domains. Future research on smart transportation, such as electric scooters, which are a great alternative to gas-fueled bikes, for example, could lead to more effective and intelligent urbanisation. Deep learning algorithms and IoT approaches for smart mobility, route optimisation, car parks, and hazard prevention have also proven to be the most common researchers have done. Nevertheless, in the context of smart transportation application issues, this analysis has found similar topics of interest, such as environmental measures, mobility finance, and people health.

Based on Fig. 11.4, this chapter has summarized two things: the least explored areas and the most explored areas in the smart city concept. The least explored research area in smart cities is smart healthcare, followed by air quality systems. However, although these two are the least to be said, a more thorough study of this area of research should be done in the future as these two are essential services, especially in urban areas and should not be neglected. Meanwhile, smart cities' most spoken research areas include smart services and smart transportation. Smart services, such as energy efficiency, water quality and trash management systems, have already been widely implemented, together with smart transportation. Therefore, this should be continued more widely as this field is vital in urban development.

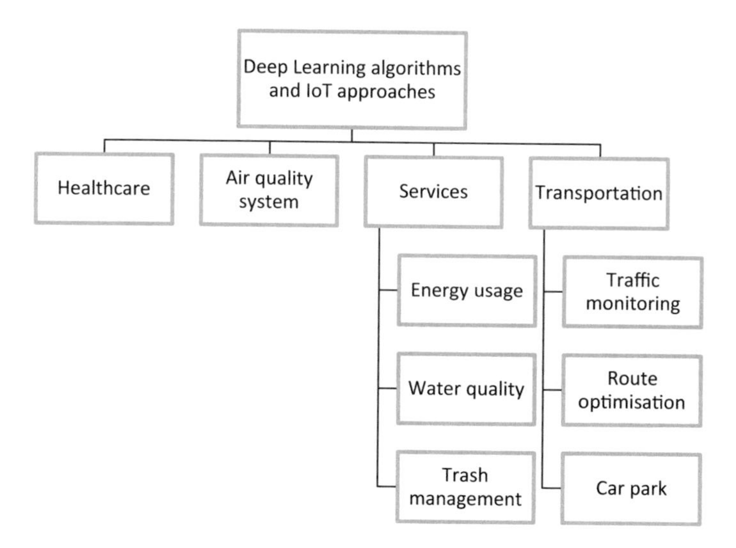

Fig. 11.4 Scope of interest for future directions of deep learning

11.4 Conclusion

This chapter reviews the deep learning techniques in artificial intelligent applications to multiple smart city domains that have been discussed, including smart transportation, smart services, smart governance, environment, security, and public safety. Several difficulties with implementing deep learning on smart city data have also been discussed. This chapter also concluded some of the deep learning techniques to help better understand the smart city concept and follow the current trends in smart cities. Smart cities have been discussed in much more detail through theory and ideologically in health and science research, bringing new industrial and academic subjects, creative and critical innovations of technologies, operations and services, and debates from multiple domains. Cities in urban areas are usually described as smart cities because these cities are more exposed to various smart applications and services. The reason is that several technologies like DL, IoT and big data have been introduced in those cities and are advancing rapidly. According to the outcome of this chapter review, these applications and services are implemented to address urban challenges and increase the standard of living. This collaboration is an excellent strategy for future smart cities to ensure automated and proficient operations.

The primary concept of smart cities is that they compromise big data analytics and IoT technologies in everyday operations. To develop and maintain smart cities, in general, energy and environmental control are among the issues that are completely important to ensure the continuous evolution of smart cities. By bringing together diverse evidence from the scientific journals to provide a more structured, systematic, and integrated perspective of smart cities, this review adds to the knowledge of smart cities. This review aims to provide an ongoing debate concerning the value of smart cities for general well-being, including economic growth and human welfare. It may highlight and justify this review by widening the transdisciplinary information of smart cities study domains.

References

1. Swan, M.: Sensor mania! The internet of things, wearable computing, objective metrics, and the quantified self-2.0. J. Sens. Actuator. Netw. **1**, 217–253 (2012)
2. Yalta, N., Nakadai, K., Ogata, T.: Sound source localization using deep learning models. J. Robot. Mechatron. **29**, 37–48 (2017)
3. Bresnick, J.: What is deep learning and how will it change healthcare? HealthITAnalytics. https://healthitanalytics.com/features/what-is-deep-learning-and-how-will-it-change-healthcare (2019)
4. Gubbi, J., Buyya, R., Marusic, S., Palaniswami, M.: Internet of Things (IoT): a vision, architectural elements, and future directions. Future Gener. Comput. Syst. **29**, 1645–1660 (2013)
5. Warif, N.B.A., Azman, M.I.S.S., Ismail., N.-S.N., Remli, M.A.: IoT-based Smart parking system using android application. In: Emerging Technology in Computing, Communication and Electronics (ETCCE), pp. 1–6 (2020)

6. Talari, S., Shafie-khah, M., Siano, P., Loia, V., Tommasetti, A., Catalao, J.P.S.: A review of smart cities based on the internet of things concept. Energies **10**, 1–23 (2017)
7. Chen, S.Y., Lai, C.F., Huang, Y.M., Jeng, Y.L.: Intelligent home-appliance recognition over IoT cloud network. In: Proceedings of the 9th International Wireless Communications and Mobile Computing Conference (IWCMC), Sardinia, Italy, pp. 639–643 (2013)
8. Ye, X., Huang, J.: A framework for cloud-based smart home. In: Proceedings of the International Conference on Computer Science and Network Technology, Harbin, China, vol. 2, pp. 894–897 (2011)
9. Martirano, L.: A smart lighting control to save energy. In: Proceedings of the 6th IEEE International Conference on Intelligent Data Acquisition and Advanced Computing Systems, Prague, Czech Republic, vol. 1, pp. 132–138 (2011)
10. Castro, M., Jara, A.J., Skarmeta, A.F.G.: Smart lighting solutions for smart cities. In: Proceedings of the 27th International Conference on Advanced Information Networking and Applications Workshops, Barcelona, Spain, pp. 1374–1379 (2013)
11. Botta, A., Donato, W., Persico, V., Pescapé, A.: Integration of cloud computing and Internet of Things: a survey. Future Gener. Comput. Syst. **56**, 684–700 (2016)
12. Gassmann, O., Böhm, J., Palmié, M.: Smart Cities: Introducing Digital Innovation to Cities. Emerald Publishing (2019)
13. Iwendi, C., Maddikunta, P.K.R., Gadekallu, T.R., Lakshmanna, K., Bashir, A.K., Piran, M.J.: A metaheuristic optimization approach for energy efficiency in the IoT networks. Softw: Pract Experience (2020)
14. Reddy, G.T., Reddy, M.P.K., Lakshmanna, K., Kaluri, R., Rajput, D.S., Srivastava, G., Baker, T.: Analysis of dimensionality reduction techniques on big data. IEEE Access **8**, 54776–54788 (2020)
15. Bhattacharya, S., Somayaji, S.R.K., Gadekallu, T.R., Alazab, M., Maddikunta, P.K.R.: A review on deep learning for future smart cities. Internet Technol Lett (2020)
16. Syed, A.S., Sierra-Sosa, D., Kumar, A., Elmaghraby, A.: IoT in smart cities: a survey of technologies. practices and challenges. In: Smart Cities vol. 4, pp. 429–475 (2021)
17. Ritchie, H., Roser, M.: Urbanization. Our World in Data. https://ourworldindata.org/urbanization (2018)
18. Ahvenniemi, H., Huovila, A., Pinto-Sepa, I., Airaksinen, M.: What are the differences between sustainable and smart cities? Cities **60**, 234–245 (2017)
19. Chen, Q., Wang, W., Wu, F., De, S., Wang, R., Zhang, B., Huang, X.: A Survey on an emerging area: deep learning for smart city data, vol. 3, pp. 392–410 (2019)
20. Ramesh, K., Lakshna, A., Renjith, P.N., Sheema, D.: Smart traffic prediction and congestion reduction in smart cities, vol. 12, pp. 1027–1033 (2021)
21. Amato, G., Inzunza, H., Robles, L.H., Mancilla, M.A.C., Neri E.L.: Traffic prediction architecture based on machine learning approach for smart cities, vol. 11, pp. 22–33 (2020)
22. An, C., Wu, C.: Traffic big data assisted V2X communications toward smart transportation, vol. 26, pp. 1601–1610 (2020)
23. Marco, C., Carrara, F., Falchi, F., Gennaro, C., Meghini, C., Vairo C.: Deep learning for decentralized parking lot occupancy detection, vol. 72, pp. 327–334 (2017)
24. Vinayakumar, R., Alazab, M., Sriram, S., Quoc-Vie, P., Soman, K.P., Simran, K.: A visualized botnet detection system based deep learning for the Internet of Things networks of smart cities, vol. 56, pp. 4436–4456 (2020)
25. Zekic-Susac, M., Mitrović, S., Has, A.: Machine learning based system for managing energy efficiency of public sector as an approach towards smart cities. Int. J. Inf. Manag. **58**, 102074 (2021)
26. Kok, I., Simsek, M.U., Ozdemir S.: A deep learning model for air quality prediction in smart cities. vol. 1, pp. 1983–1990 (2017)
27. Aqib, M., Mehmood, R., Albeshri, A., Alzahrani, A.: Disaster management in smart cities by forecasting traffic plan using deep learning and GPUs, vol. 224, pp. 139–154 (2017)
28. Kinoshita, S., Wakatsuki, M., Horinouchi, S., Tanabe, S., Higashi, A.: Fujitsu's deep learning technology that enables smart city monitoring, vol. 55, pp. 30–37 (2019)

29. Vandecasteele, I., Baranzelli, C., Siragusa, A., Aurambout, J.P., Alberti, V., Alonso Raposo, M., Attardo, C., Auteri, D., Barranco, R., Batista e Silva, F.: The Future of Cities–Opportunities, Challenges and the Way Forward. Publications Office, Luxembourg (2019)
30. Kang, M., Choi, Y., Kim, J., Lee, K., Lee, S., Park, I.K.: COVID-19 impact on city and region: what's next after lockdown? Int. J. Urban Sci. **24**, 297–315 (2020)

Chapter 12
Smart and Sustainable Cities in Collaboration with *IoT*: The Singapore Success Case

Roberto Ferro-Escobar, Harold Vacca-González, and Harvey Gómez-Castillo

Abstract A smart city is an urban space focused on improving the quality of life of its citizens through the intensive use of technology to support social and economic development, where multiple areas cooperate systematically to achieve sustainable collective results at all levels. In this sense, Internet usage has reached a point where it has become essential in everyday life: the need to have information at hand in the shortest possible time has generated a technological revolution around the constant connection with this tool. Thus, personal life and everyday objects have created the need to continuously monitor and understand what surrounds the human being, hence the birth of the Internet of Things (*IoT*); and the convergence between the environment, human interaction and technology has given rise to the concept of smart and sustainable cities (*SSC*). However, given the current state of these concepts, it is necessary to apply for an introductory contextualized conceptual review on Smart and Sustainable Cities, and then characterize them from the establishment of broad categories with their respective subcategories: Smart Mobility (Mobility Integration, Sustainable Public Transport Service, Electric and Autonomous Vehicles); Energy Efficiency (Renewable Energies, Natural Resources Management and measurement of environmental parameters—Emissions, Water Consumption Efficiency); and Smart Health (telehealth and solutions provided by information and communication technologies (*ICT*); and the use of *ICT* for generation, transmission, and processing of information (Machine learning, *IoT*, sensor grids, Cloud Computing, Big Data). Based on the above, this chapter comparatively addresses concepts on the present of *SSC* in collaboration with *IoT*: advantages and disadvantages, reception, adaptability, and vulnerability in data. The city-nation of Singapore is assessed through a quantitative evaluation in terms of *GMSDIL* (Governance,

R. Ferro-Escobar · H. Vacca-González (✉) · H. Gómez-Castillo
Universidad Distrital Francisco José de Caldas, Bogotá, Colombia
e-mail: hvacca@udistrital.edu.co

R. Ferro-Escobar
e-mail: rferro@udistrital.edu.co

H. Gómez-Castillo
e-mail: hagomezca@correo.udistrital.edu.co

© The Author(s), under exclusive license to Springer Nature Switzerland AG 2022

G. Marques et al. (eds.), *Machine Learning for Smart Environments/Cities*,
Intelligent Systems Reference Library 121,
https://doi.org/10.1007/978-3-030-97516-6_12

Mobility, Sustainability, Economic Development, Intellectual Capital and Quality of Life) as a successful city, concluding from the qualitative description under this model, an urban resilience that has allowed the improvement of the quality of life of its citizens, even in traumatic circumstances and catastrophic effects such as those of the current pandemic.

Keywords ICT · Internet · IoT · Smart and sustainable cities · Urban resilience · Singapore

12.1 Introduction

The historical trend towards urbanization has been an essential feature of the structure of capitalism since the first industrial revolution. As a result, throughout the different processes of rural–urban migration, the latter has been configured as an urban geographic space where multiple social, political, economic, and cultural dynamics are articulated. It is there where, since the twentieth century, technology has been incorporated into city planning and management; in the last two decades, *information and communication technologies* (*ICT*) have been used intensively, creating narratives that promote economic development, the improvement of the quality of life of the population and the future prosperity of cities. Under this slogan, cities have been built in the *twenty-first* century from scratch, where the principles of intelligence are provided by *ICT*.

Consequently, the inclusion of technology in the dynamics of cities has generated a series of discourses that defines it as future utopias, in two ways.

The first of them proposes a prediction, the disappearance of cities. Goldmark in [1] suggested the use of communication technologies to consolidate new rural societies that could counteract the demographic explosion and the urban and suburban life trend, since "*60% of the population lived on 10% of the land*". Likewise, the utopia proposed by the North American urban planner Frank Lloyd Wrigh [2, 3], called the "*broadacre city*" -a theoretical design based on decentralization that guaranteed the separation of cities where the factories, farms, and houses could coexist side by side [4]- was debated and reevaluated. Jane Jacobs [5], in contrast, would affirm that the massive use of automobiles in cities—an opinion shared by urban planners with their modernizing tendencies and processes of urban reconstruction and those representing economic interests- would end up destroying the fundamental social elements of life in cities. In the same direction, M. McLuhan would argue that electronic media would eliminate geographical barriers and consequently the dissolution of cities, [6].

The second, proposes a vision that considers that technology solves -by itself- all city problems by incorporating science and technology to build modern and healthy conditions; a futuristic vision where *ICT* are included in the dynamics of urban development in the twenty-first century as the backbone of growth, efficiency and prosperity of future cities, [7].

In such discourses, *Smart Cities* (*SC*) would become the panacea that would provide solutions to the problems inherent to the growth of cities, promising a future that symbolizes a new type of urban utopia driven by technology [8, 9].

Based on the above, and beyond the utopian visions of city-technology relations, there is a perspective that places technology as a catalyst and dynamizer of economic activity. This symbiosis reconfigures both physical spaces of production, the network of financial markets, and in general, the city's infrastructure. Although today's cities go beyond these changes, they constitute "*a huge network of cyber-connections focused on optimizing the consumption of urban resources and processes of prevention of negative external effects, resulting from city operation*" [10], p. 138. Technologies that support the globalized network require robust physical infrastructure strategically located in cities, giving way to what is called "*Global Cities*" [11]: a reduced number of cities that not only concentrates the main functions of the world economy but also groups the infrastructure represented in ports, airports, railways, roads, subways, suburban trains, money and data flow in the global network; precisely, this is how technology impacts urban space: it transforms the economic environment, which in turn changes the way the territory is sustainably colonized [12]. It is in this context that the so-called smart and sustainable cities *SSC* are framed.

In this sense, the evolution of the Internet has catalyzed the condensation of massive development of devices that allow constant cyber-connection with the environment and, at the same time, allows the transmission of this information to human beings, giving them an effective sense of control over the things that surround them. This is how concepts such as the *Internet of Things* (*IoT*) have achieved a significant increase in such relationships, both in domestic and productive life, which can reach even the largest cities, as well as the evolution of certain cities through such principles [13, 14].

However, to the technological framework indicated, which has opened up an infinite number of relationships and possibilities, it is necessary to maintain an anthropological sense expressed in the fact that citizens are called upon to innovate in order to actively participate and adapt, beyond considering them as simple urban sensors whose role is subaltern in scenarios of social dynamics, to the disruptive changes brought about by new cultural demands generated, for example, during the pandemic, [15]. This is one of the main contributions of the chapter: the *urban resilience* of *SSC* must be increased collectively because in addition to technological considerations, the capacity of citizens to play a role that is not exclusively individual or at the margin of collective aspirations.

As a result, specific objectives of the chapter are set out as follows. In the first part, the focus will be on the couple that will lead to all the topics discussed: the collaboration between *IoT* and *SSC*; with different approaches that allow to understand the joint work that they have achieved in a subtle way with an indirect but deep relationship; its advantages and disadvantages compared to the most important points of view of its use; the disclosure that this has had for its understanding in multiple areas concerning the services and utilities that will provide in different areas of both knowledge, professional, labor; and, from this comparative, observe how these features have been coupled in everyday life, both at personal, family, and in the city, [16, 17].

The second part will focus on the impact of technological development and the progress of interconnectivity and interoperability that objects have gained with people through the Internet at a social, economic, and cultural level. This topic will be treated from different points of view: as the development that it has had, from a personal level, the corporate usability; as well as its implementation in cities, emphasizing the background of all these advances; and, beyond, the impacts, influences, and consequences that this has had on society at economic and social levels. This last to cover a larger topic, the academic approach: expansion of new professional possibilities, not only technically but with a global vision, of the effect that people with knowledge in related areas of technological studies can have in today's world, [18, 19].

Finally, it is intended to demonstrate the current status of these concepts globally through advances in devices designed to assist in the development of projects, such as specialized hardware, e.g. prototypes and current marketable products at the forefront of this market; projects and applications that are currently being developed. Everything to highlight the progress that our duo has, showing the *nation-city* of Singapore as a successful case because of its efforts in the implementation of ideas and projects that have allowed it to stand out as one of the Smart Cities established in the world.

The chapter is organized as follows: initially the background related to the conceptualization of *SSC* in relation to the *IoT* is presented; then methodological considerations that have led to the classification of *SSC* are made; subsequently, and as a result of an evaluation of Singapore as a success story, these results and their discussion in terms of Governance, Mobility, Sustainability, Economic Development, Intellectual Capital and Quality of Life are presented; finally conclusions are drawn and future scopes.

12.2 Background Related to Smart Cities in Perspective of *IoT* Collaboration

The term "*smart city*" is related to a series of neologisms coming from the academic world that reflect changes occurring in the urban space with the inclusion of *ICT*, such as: Informational City [20], Technopolis [21], Metapolis [22], Postmodern City [23], Virtual City [24], Postmetropolis [25, 26]. However, the notion that echoed in the world of marketing and urban management policies in the last two decades of the twenty-first century was the *smart city*.

Regarding the origin of the term, for some authors such as Harris [27], the closest antecedent is found in the work of Sheridan Tatsuno, referring to transformations in traditional cities and high-tech technology parks—with an advanced incorporation of information and communication networks—where the information management support is provided by processors [28]. However, the first works that make use of the term *SC* are those of Batty [29] and Laterasse [30] at the beginning of the nineties of the previous century, a term that began its expansion in praxis as a development agenda for worldwide cities since 2009 [7].

Regarding *Smart City* conceptualization, among many other considerations that can be found in the existing literature, two basic perspectives will be adopted.

The first of these is the technical perspective that focuses *SC* on *ICT*, giving it a trait of intelligence to the technology and software that is incorporated into the city [31]. In this sense, the purpose of *ICT* deployment is to monitor, measure, manage, operate and make more efficient decisions from the public administration of cities to interconnect inhabitants with institutions and the infrastructure of public services, under the premise that these technologies act in a *"smart way"* [7, 32]. In speeches and marketing strategies of global technology corporations such as IBM—promoted in the last decade—the use of *ICT* in city governance is identified, giving relevance to the data permanently produced by these systems in preventing and solving problems, affirming that it has an efficient impact on the use of resources, making cities more attractive places in the twenty-first century: both for investment and for citizens with the economic capacity to live in demanding cities.

But the above is not only typical of corporations but also in academic discourses. W. Mitchell attributes the intelligence of cities to the efficacy of networks, data processing, sensors, and support software tools, [33].

In the same perspective, network infrastructures and Internet-based applications have been used in urban policy for urban spatial planning and management under the premise of improving the management of governments and the quality of life of its inhabitants [34]. With massive amounts of digital data and the use of wireless technologies with higher performance, reliability, security, and real-time operation, there has been "*a new understanding of urban problems, effective and feasible ways to coordinate urban technologies; models and methods for using urban data at spatial and temporal scales*" [35], p. 481; laying the foundations for a new, highly rationalized techno-scientific implementation process for city planning and management [32], characterized by highly specialized planning, such as that corresponding to transportation networks—better known as *smart growth*—whose objective is to avoid cost increases caused by cities' territorial expansion [36].

Furthermore, city planning and management have brought a future vision that considers *ICT* as a solution to the city's economic problems [8]. Economics provides positive statistical evidence between *SC* policies and economic growth [37], as well as the relationship of teleinformatics infrastructure and improvement of the economic performance of cities [38], elements that lead to the high acceptance of *SC* in their governmental policies.

Therefore, in terms of economic globalization, *SCs* acquire an extra value that operates in two ways: from the outside, making them visible as attractive cities for foreign investment, and equally attractive for the new labor force represented by highly qualified specialists who constitute the new *"creative class"* [39]. In this way, governments have tried to take advantage of *ICT* in the development of innovation networks to boost the economy, promote progress, and make cities more competitive; in other words, the success of policies must be evaluated along with these guidelines. In a second direction, a cost–benefit culture between taxpayers and the uses of goods and services is strengthened [40], p. 6. The consequence of assuming this perspective is unquestionable: the future of the world's cities will exclusively rely on embracing the technological impulse provided by *ICT*.

The second approach is the sustainable one. From a wide scope, the notion of sustainability includes the improvement of the quality of life of people and communities living in cities in the long term; quality of life expressed in terms of comfort, safety, and savings. City sustainability would imply at least four aspects of sustainable economic growth: quality of life, participatory governance, and environmental sustainability [41]. In this direction, the role of *ICT* and more advanced technologies that have been developed in the last decade should allow the creation of jobs in the new service economy [42], make urban planning and management more efficient, and thus strengthen freedom of expression, public information, services and citizen participation in bureaucratic decision making. Nevertheless, the most notable aspect of this perspective has to do with environmental sustainability aimed at reducing energy consumption and environmental footprint in the idea of consolidating ecological or green cities [43]; ensuring that the use of innovative and integrated state-of-the-art technologies contributes to the reduction of carbon emissions or the efficient use of energy through renewable energy sources, are purposes that must be evaluated in *SC* management [44].

Consequently, from this sustainable perspective, a more comprehensive Smart City posture emerges aimed at balancing economic, social, and environmental sustainability with the use of *ICT*, in an attempt to increase the quality of life of the population [40] called: sustainable smart city or *SSC*, which, depending on the type of approach, emphasizes some factors more than others [23]. It highlights, beyond the technological infrastructure represented in digital telecommunications networks, advanced technologies in services, to make management more effective, oriented to the theory of sustainable development and relevant to the development of human and social capital. Along these lines, the city should be considered a space for learning, innovation, and creation, with institutions that support research and progress, interconnected with higher education institutions in search of capital creation. Along the same route, human capital, cities must be built to prevent social inequality by guaranteeing social and digital inclusion [45], promoting citizens with the capacity to incorporate *ICT* innovations that also generate a greater sense of belonging to the city [46].

As a result of the above, some international organizations sponsored by the United Nations have taken on the task of promoting this type of city in Europe - through public policy-. Within this framework, a denomination developed by the United Nations Economic Commission for Europe *(UNECE)* and the International Telecommunication Union *(ITU)*, which has gained relevance and acceptance this past year:

> A smart sustainable city is an innovative city that uses information and communication technologies (ICT) and other means to improve quality of life, the efficiency of urban operation and services, and competitiveness, while ensuring that it meets the needs of present and future generations concerning economic, social, and environmental aspects as well as cultural aspects, [47].

To establish the ideal convergence between *IoT* and *SSC*, however, it is necessary to discuss some points of view regarding its development and implementation, as well as how this convergence has grown up, [48, 49].

Following this last point, it will be relevant to describe advances in devices applied and incorporated in *SC*: specialized hardware, prototypes, and marketable products at the forefront of the market, model projects and applications, among many elements that, when scaled, may have an impact on emerging experiences.

IoT and *SSC* have relevant aspects that have helped both improvement, optimization, and performance in many of the areas where they have been deployed, contributing to the solution of complex problems. Precisely, detecting them in an agile manner is one of the main benefits of introducing technological solutions. Today, the verification process is performed by a personal and manual visualization, but with an interconnected system, faults appear immediately on screen, which allows a more efficient repair; thus, the most important thing is to be able to monitor their status to proactively plan their maintenance and minimize them, being this one of the main advantages when opting for this implementation, [50, 51].

IoT has allowed, on the other hand, the integration of people, objects or things, data, and processes, which are fundamental pillars in the development and evolution of technological infrastructure through enhanced interoperability and cyber connection of different areas, [52, 53]. Its implementation has achieved time savings by minimizing monitoring, management, and supervision tasks. Likewise, the models and analysis of massive and fast data provide correct decision making and the immediate detection of failures that arise in supervision and personal management, the monitoring of processes through computers -as well as the number of objects or processes carried out simultaneously-, and resource administration at an economic level, among others, [54, 55].

Also, by embracing a complex infrastructure for monitoring or managing processes, *SSC* optimizes performance and enables the integration of processes in public administration. In this sense: it provides necessary and transparent information for better decision making and budget management; it generates common procedures that increase the efficiency of government or city administration; it optimizes the allocation of resources and helps reduce unnecessary expenses; it raises satisfaction levels among inhabitants since its ideal is to provide better attention to users and to improve the image of public entities. In other words, it produces performance indicators that are useful for measuring, comparing, and improving public policies, [56, 57].

However, like any new development, in the case of *IoT* and *SSC*, there has been a social impact of technological advances in aspects that, due to their recent arrival, are challenging and require continuous improvement.

Therefore, as *IoT* is based on the interconnection of everything that surrounds people through sensors and actuators; to have quality control over them, it is necessary to develop standards for labeling and monitoring, [58]. It is necessary to create them: i.e., for the universal serial bus *(USB)* and Bluetooth, there are multiple risks concerning the handling of information, as accidents can happen regarding device responsibility level.

Security and privacy, besides, is another questioned and debated problem regarding the massive use and implementation of *IoT*. As the possibility of cyber-attacks and personal data leaks is present, chances are endless and all security risks become the responsibility of users, [59].

Among other factors that ease the convergence of *IoT* with *SSC* is the access and quality of the Internet. The network of networks makes it possible to interconnect all objects, achieving interoperability and efficient management of processes; intelligent environments adapted employing *ICT* are established on it.

Another element that manages to establish convergence in this duality is sensor management. For *IoT* and *SSC* it is essential to develop their purposes; here the important thing is that each of the measurable variables that surround a specific environment can be quantified, but thinking and moving to a domestic scale: gas or temperature meters; or to a metropolitan scale: electricity consumption measurement in street lighting, or motion measurement for traffic detection, [60].

Furthermore, the delegation of small-scale processes -such as placing orders for missing groceries in a house-, up to predictive and corrective maintenance -such as that required by large-scale automatic industrial machinery- involves supervision that, although it is in line with the progress and development of the aforementioned concepts, must be constant, since the risk of failure or error is critical. Thus, it is necessary to have a minimum knowledge of these technologies, taking into account their scope as well as their limits. It is a factor that does not change regardless of the responsibility given to the devices or their scale of implementation, [61].

Reinforcing that an *SSC* places people at the center of development, the incorporation of *ICT* in urban management uses its elements as tools to stimulate the construction of an efficient government that includes collaborative planning processes and citizen participation. Being the *IoT* a concept that refers to digitally interconnecting everyday objects with the Internet, its relationship with the need to solve problems and continuous improvement with different perspectives regarding the scale of operation, it must have a social approach that stands out over the correct technical management, so that responsibilities must be assumed in all areas: homes, companies, or the city in general, [62, 63].

Therefore, it should be taken into account both public and private organizations' responsibility to become articulators to contribute to the development of smart cities through security solutions that generate greater wellbeing for citizens. In this context, the implementation of technologies aimed at generating large volumes of data must be managed by expert personnel, making it imperative to have a critical and operational mass in the institutions that run these processes to handle information in a suitable way, [64, 65].

For *SSC*, all the alternatives that are given above can be applied to public building management (*inmotics*); as well as the maintenance of public infrastructures: electrical networks, water supply pipes, lighting, sewerage, traffic management elements, screens, parking areas, irrigation, parks, and gardens, among many others.

Along the proposed path, and given that buildings are the basic components of the cities -which consume about 40% of all the world's energy, and it is estimated that 50% of this consumption is not efficient (in the case of the United States, buildings consume 70% of all electricity, of which 50% is wasted, while 50% of the water consumed is wasted)—it is reasonable that applying technology to improve this management should be one of the most pressing areas for the *SSC*, [66–68].

There are many alternatives, but the key lies in building a single network. This process allows the automation of lighting operation to overcome the challenges of traditional or analog lighting so that the sensors involved in the network can detect the levels of natural light around the luminaire allowing to automate its switching ON and OFF, as required. On the other hand, also by centralized action of the operator, the power of each lamp can be programmed individually to lower its consumption and thus create lighting profiles for streets, sectors, or schedules, as appropriate. The latter is a simple example of the many that can be developed in an *SSC*.

According to this scenario, citizen surveillance projects are those that have made their way, although in a dispersed way: installation of cameras in strategic points of the metropolis that are activated in case of emergency, reinforcement of population security, systematization of the measurement and monitoring of mobility in the city through the installation of technological platforms that provide results in real-time for decision making, [69–71]. In this sense, the needs of cities in terms of security are opportunities for *ICT* to contribute to their transformation by seeking to establish smarter and safer places.

However, even though the handling of data generated in an *SSC* by the thousands of connected sensors tends to improve, it is still a step behind what it is designed to protect. The Internet was developed one way and used another. It did not occur to anyone that the security requirement was going to be critical, and the problems were only patched, [72]. Now, program designers and developers, as in electronic device applications, are more aware of the situation so that the security problem is beginning to be addressed comprehensively across all components. However, there are still applications with vulnerabilities and devices whose communication protocol is secure but not understood that way by the user. For example, there is certain information that should not be revealed or programs that should not be installed because something may be forgotten due to manual activity, [73]. In addition to the above, there is the introduction of errors in the code of an application or its protocols, and there is still no efficient technology to evaluate in an automated way the dynamic detection and incorporation of reaction mechanisms to unknown events. In this scenario, actions are very limited; although there are checklists and some tools for evaluation and exhaustive software testing, [73].

Furthermore, to speak of adaptability, a concept of service before that of *SC*, implies making necessary reference to services related to city government and its relationship with citizens in terms of transparency and participation in decision-making. In this perspective, the implementation of educational programs on the proper use of technologies is opening the way to promote their appropriation among communities through inclusive systems. Such initiatives have resulted in rules, regulations, laws, and policies for technological development, which, within a legal framework, help to promote a culture of change in cities [74].

In addition, there are online participation services and initiatives that, in general, promote transparency and contribute to the governance of an *SSC*. Among the most common implementations are sites for conducting surveys and voting, as well as social networks that encourage communication and association of different *stakeholders* to generate a constant integration and relationship with existing and implemented technologies, [75].

12.2.1 A Look at Devices

Development models and improvement and management processes have created multiple alternatives for problem-solving focused on the use of smart devices for *SSC*, among them: *Smart Mobility, Smart Industry, Smart Security, Smart Grids, Smart Energy, Smart Environment, Smart Health, Smart Home, Smart Retail, Smart Buildings, Smart Metering, Smart Street Lights, Smart Parkings, Smart People, or Smart Governance.*

In terms of devices, it is worth highlighting the topic of citizen mobility, an issue that causes permanent conflict at social, environmental, and traffic levels. It has been recognized some deployment of electric vehicles in cities and other motorized elements, and its respective public and private recharging stations, which fall into the so-called *Smart Mobility*—Tesla cars, of Elon Musk, for example, [76].

On the other hand, devices of interconnected intelligent networks, also known as *Smart Grids*, can guarantee the circulation of data in both directions—between the service center or control center and the citizen. But this agility of massive data traffic requires other models and devices that regulate such data in a better way, [77].

The other element involves the in-building devices in an *SSC*. Domotics, among other environmentally friendly electronic alternatives that have integrated energy and consumption monitoring systems, known as *Smart Building*, has shown significant improvements with robust implementations, [78].

In addition to the three previous models, there is the so-called *Smart Metering*, or intelligent measurement of public services and consumption costs, which provides higher quality data on energy consumption for each user. For example, using remote meters installed mainly in electric utilities, readings are taken remotely and in real-time.

In any case, it is the sensors provided with intelligence that will have the function of compiling every one of accurate data for an *SSC*. They are an essential part to sustain connectivity and information, and make each subsystem fulfill its function, [79].

An additional model to the standard *SC* model, where citizens and inhabitants are certainly the essential part of the city since without their active participation it is not possible to realize this abstraction, called *Smart People*, can be seen [80].

Therefore, the growth of devices and specialized hardware for the development and implementation of devices involved in *IoT* (air quality, water quality, gas, and water leak detection, electromagnetic emissions detection, among the most important ones), is becoming more and more available to people, since a knowledge base of relative ease to design and program from scratch a project with a professional profile has spread.

Some typical simple hardware items may include: Arduino, boards able to read inputs from a sensor and convert them to an output; Raspberry Pi, small board computer or single board computer (SBC); Portable Wireless N router, capable of sharing a 3G/4G/5G mobile connection, compatible with 3G/4GLTE/HSPA + /UMTS/EVDO USB modems; ESP8266 Wi-Fi module, which allows microcontrollers to connect to a Wi-Fi network and make TCP/IP connections using Hayes-type commands ACS720 current sensor, a Hall-effect sensor that detects magnetic field by induction; LM393 sound sensor, which controls lights, alarms, and sound tracking robot through a potentiometer that allows setting the volume; MQ-135 air quality sensor, in buildings/homes detects the presence of NH_3, NOx, alcohol, benzene, smoke, CO_2; DHT11 sensor for humidity and temperature, consisting of resistive sensors with fast response in humidity measurements in the range 20–95%, and temperatures in the range 0–50ºC, [81–85].

Other, more sophisticated, high-sensitivity sensors have been developed to detect, i.e., electromagnetic emissions up to 20 GHz, as well as weak environmental signals such as FM, AM, Wi-Fi, and Bluetooth radio, through spectrum analyzers; this opens up unusual possibilities to monitor multiple variables, including ambient temperature, much more effectively than standard radio frequency receivers and analyzers [86]. In this perspective, research, development, and innovation have not stopped.

12.3 Methodology for Classifying a City as an *SSC*

In practice, *SSC* have been measured by indicators that relate to categories of the ecological economy and bio-economy concepts. Likewise, rankings have been established for evaluating the indicators to identify its appropriation by the different cities of the world that pursue such classification. According to the second edition of the IMD-SUTD 2020 smart cities report [87], in an international context, more than a hundred cities are typified as *SSC*, where some pioneering cities -built from scratch by multinational technology corporations- do not appear in this ranking, but they are interesting cases, nonetheless.

Songdo—in South Korea—and Masdar—in the United Arab Emirates—stand out for this characteristic. They are *SC* built from scratch, guided by *"technological optimism"* and the technological perspective that focuses the *SSC* on *ICT*. The fact that they are new cities makes it possible to remove all the problems of excessive growth of traditional cities, disorderly urban growth, limited infrastructure, and overflowing population growth, which leads to very strong demand for services.

Songdo, a city built in 2015 (although the first inhabitants settled in 2008), with a central park and a large canal extension (inspired by New York and Venice). It was designed by the architectural firm Kohn Pedersen Fox (*KPF*), with advanced infrastructure by the U.S. firm Gale International, the Korean firm *POSCO E&C* and the public entity of the metropolitan city of Incheon [88], while the technological development was carried out by *CISCO*. It has a technological infrastructure located at the city core, in charge of the operations control system; based on *IoT*, *RFID/USN*

(Radio Frequency Identification/Ubiquitous Sensor Network) or Transferring of data and power using contact-less technologies: they monitor and control transportation, security, disasters, environment and citizen interaction, as well as homes, businesses, education, health, and traffic. The city aims to become a knowledge and information industrial complex with *ICT* facilities, biotechnology, and research and development centers [89]; conditions that have conferred it the title of "*ubiquitous city*" [90].

Masdar, on the other hand, has been the city that generated the highest expectations in its marketing process. In 2007 it was promoted as the city that aspired to be the world's first carbon-free city and the most sustainable city on Earth. By 2012 it was promoted as the world's first clean-tech, sustainable, renewable, and energy-powered cluster [32]. Some of the features that can be highlighted have to do with the city's design that merges cutting-edge technologies with traditional Arab architecture. The power source is based on solar energy, the university—which is the emblem of the city and guided in the early years by *MIT* (Massachusetts Institute of Technology)—is now a center of higher studies in artificial intelligence [90]. Masdar's *SC* experience, however, has been labeled as a technological failure, due to its inability to fulfill the expectations generated [91, 92]. In any case, what is certain is that *SC* built from scratch leave some questions in the academic community; and Masdar is no exception: the process of co-modernization based on high technology is considered to be guided more by market criteria than by ecological studies [93].

As indicated, the case chosen to classify an *SSC* corresponds to Singapore: it has progressively managed different systems in the city to continuously improve, from *ICT* infrastructure and sustainability, social and economic aspects to raise the quality of life of its citizens.

It is for this reason that the success case chosen belongs to Singapore. Hence, the construction of the initial existing relationships between systems and elements that Singapore has incorporated to become an *SSC* is almost natural and leads to an Integrated and Complementary Network of Systems (*ICNS*). Figure 12.1 shows this systemic integration and complementarity. The model -mind map- depicted, sets out -from the given domains- how, through the use of technology, it is possible to address citizens' problems to provide effective solutions. The *IoT* -fused with *Big Data* and data management models- makes it possible to innovate and adapt to effectively manage the city.

From *ICNS,* the *GMSDIV* (Governance, Mobility, Sustainability, Economic Development, Intellectual Capital, Quality of Life) model given in [94] is adapted and used to evaluate the city of Singapore.

The model is presented as follows: Table 12.1 describes the axes and corresponding factors that are taken into account to evaluate a city as a smart city. Each factor is evaluated on a scale from 0 to 4 and the factors of each axis are evaluated as follows:

$$\mathrm{G} = (\mathrm{g1} + \mathrm{g2} + \mathrm{g3} + \mathrm{g4})/4 \tag{12.1}$$

$$\mathrm{M} = (\mathrm{m1} + \mathrm{m2} + \mathrm{m3} + \mathrm{m4})/4 \tag{12.2}$$

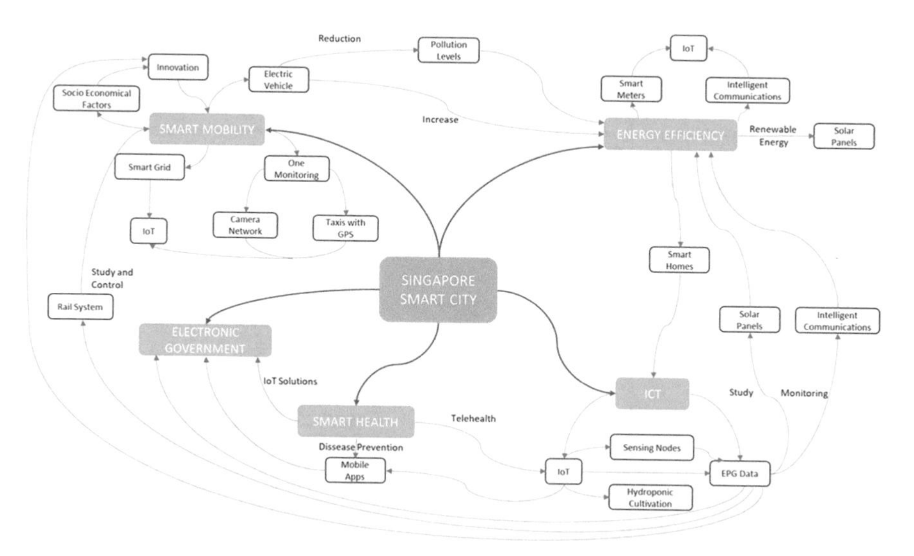

Fig. 12.1 Integrated and complementary network of systems and elements that make Singapore a successful case as an *SSC: Smart Mobility, Energy Efficiency, Electronic Government, Smart Health, ICT*. Its structure establishes five categories with its respective subcategories: *Government Policies and Innovation* (e-government, e-governance, directive intervention, facilitative intervention); *Smart Mobility* (Mobility Integration, Sustainable Public Transport Service, Electric Cars, Autonomous Vehicles, Transportation and Citizens); *Environmental sustainability* (Energy Efficiency, Renewable Energy, Natural Resource Management and measurement of environmental parameters); *Smart Health* (Telehealth and solutions provided by *ICT*); and the use of *ICT* for the generation, transmission, and processing of information (Machine learning, Internet of Things (*IoT*), Sensor Networks, Big Data)

$$\mathrm{S} = (\mathrm{s1} + \mathrm{s2} + \mathrm{s3})/3 \quad (12.3)$$

$$\mathrm{D} = (\mathrm{d1} + \mathrm{d2})/2 \quad (12.4)$$

$$\mathrm{I} = (\mathrm{i1} + \mathrm{i2})/2 \quad (12.5)$$

$$\mathrm{V} = (\mathrm{v1} + \mathrm{v2} + \mathrm{v3})/3 \quad (12.6)$$

$$\mathrm{SC} = (\mathrm{G} + \mathrm{M} + \mathrm{S} + \mathrm{D} + \mathrm{I} + \mathrm{V})/6 \quad (12.7)$$

- *SSC:* rated above 2
- Conventional cities: rated with 2
- Below average cities: rated below 2.

Table 12.1 *SSC* axes and evaluation factors

Smart and sustainable city axes		Factor	
G	Governance: e- Government y e- Governance	Electronic headquarters	(g1)
		Transparency	(g2)
		Interactive street map	(g3)
		Communication with citizens	(g4)
M	Mobility	Sustainable urban mobility plans	(m1)
		Multimodal integration of public transport	(m2)
		Deployment of alternative means (bicycle)	(m3)
		ICT in traffic control	(m4)
S	Sustainability	Energy efficiency	(s1)
		Water consumption efficiency	(s2)
		Emissions	(s3)
D	Economic development	Open Data	(d1)
		Innovation ecosystem	(d2)
I	Intellectual capital	Wi-Fi as a municipal service	(i1)
		Training, education	(i2)
V	Quality of life	Health and sanitation	(v1)
		Universal Access	(v2)
		Deployment of several ICT measures for the enjoyment of the city	(v3)

12.4 Results and Discussion of the Evaluation of Singapore as a Success Case

Finally, applying (7), see Table 12.2, with the values of each factor for each axis, gives a score of 2.46—rated above 2; indicating that Singapore is valued as a success case of *SSC*.

Table 12.2 Values applied to (7) on Singapore, indicating its qualification as a Smart City

G		M		S		D		I		V	
g1	3	m1	4	s1	4	d1	2	i1	1	v1	3
g2	3	m2	4	s2	4	d2	4	i2	2	v2	1
g3	1	m3	1	s3	2					v3	1
g4	2	m4	3								
	2.25		3		3.33				1.5		1.66

Score 2.46

12.4.1 Governance (Electronic Headquarters, Transparency, Interactive Street Map, Communication with Citizens)

Government plays a decisive role in a city's decision making. From this perspective, Singapore has contemplated in its smart nation plan the integration of different government entities to create an environment of mutual cooperation where the city can be managed efficiently.

Furthermore, has seen the need to develop an integrated data-sharing platform to leverage the existing smart sensor network to provide accurate data on the city's performance where collected information is processed to enable efficiency in the city's management of the systems it has to control. In 2013, *SCDF* (a government agency) aided more than 4000 fires, which represented a 7.8% decrease from 2012 cases. In 2013, as a result, it recorded the lowest annual number in 20 years. Similarly, the benefits of surveillance cameras can be measured through the number of crime cases it has helped solve and ultimately reduce the urban crime rate.

12.4.2 Mobility (Sustainable Urban Mobility Plans, Multimodal Integration of Public Transport, Deployment of Alternative Means, ICT in Traffic Control)

The *SSC* Singapore is characterized by an intelligent transportation system (*ITS*) developed in 2006 where a Mass Rapid Transit (*MRT*) system is established to provide national interconnectivity [95]. It has a monitoring system (*one monitoring*) to access traffic information that is collected by a camera network along the streets and by cabs with GPS. Likewise, the application developed by the *LTA* monitors incidents on the roads, and an intelligent electronic device (*Your speed sign*) alerts drivers about the speed limit [96–98]. E-payment to access the transportation system generates benefits that are reflected in time reductions to reach the different means of transportation, guaranteeing convenience to users, [99].

Thus, the option to integrate different means of transportation is through a shared service that allows several users to make use of the same vehicle to travel to different points around the city. Mobility concepts such as *MaaS* (Mobility as a Service) help to buy mobility services, which are packages based on consumers' needs instead of buying them directly from the means of transport. Another feature offered by this service is the integration of ticketing and payment that through the use of a smart card can access multiple modes of transportation and charging is performed based on the use of each service accessed. The *EZ-link* card is widely used in Singapore by users because it eliminated payment barriers between operators in transit networks [100–102].

On the other hand, public transportation provides tools consistent with dynamic and transformative development for *SSC* Singapore [103–105]. It is important to highlight those socioemotional factors have a significant influence on the decisions that people make to access or use some means of transport, taking into account these factors transportation modes ease the mobility flow of users [106, 107].

It is noteworthy that, in Singapore, the widespread use of electric vehicles (EVs) reduced air pollution levels compared to traditional combustion vehicles; in turn, EVs offer higher energy efficiency and lower energy costs [108, 109]. Charging stations for electric cars are controlled and monitored by an electric grid management system to establish which point is available to charge the car, avoiding congestion at the stations and making it easier to charge the vehicle owner [110–112].

Similarly, the arrival of autonomous vehicles soon will require an understanding of the benefits they provide and how they can be adapted to the existing mobility ecosystem to obtain the best results in their implementation. In Singapore, through SimMobility (agent-based integrated supply and demand model) application, shared autonomous vehicles could anticipate the demand in mobility service reducing waiting times and ensuring a balance between required vehicles and available vehicles in each area [113].

In another but complementary area, the adaptation of means of transport to the needs of all citizens is a task that government entities and involved stakeholders must carry out if the best services are to be provided in *SSC*. In [114], for example, the study of the activity patterns of people over 50 years of age (trend age) is reported to know in detail their use of time to optimize the performance of activities and thus offer the transportation methods required by this age group.

12.4.3 Sustainability

On this topic, Singapore introduced an electrical network with smart meters to know the electricity consumption and electricity supply to buildings by vehicles to infrastructure (*V2I*) [109], to perform subsequent monitoring and achieve cost reduction. Electricity consumption levels can be compromised if this electricity flow is not properly controlled, and therefore customers are indispensable to make this supply control effective in Singapore's buildings [115–117]. Defining the appropriate rules or laws that regulate these *Smart Grids* is important to normalize and ease their implementation, [118].

From the above, it was possible to improve energy efficiency in controlled environments where, seeking comfort for building residents, intelligent systems such as lighting with *LED* bulbs were designed and deployed. This kind of system was remotely controlled to set lighting parameters concerning natural light or by the number of occupants of these buildings [119, 120].

On the other hand, smart homes had another option to reduce energy consumption levels, a measure that contributed to regulate and control global warming; however, it was important for the citizen to adopt measures to reduce these energy consumption rates assisted with smart systems installed in homes. For instance, by using smart systems and sensors (Motion, temperature, humidity, light detection sensors) it was able to control household appliances or light sources in each of the spaces of the house [121, 122]. However, it was necessary to take into account the security of this kind of home due to the amount of data handled, which can be easily accessed through the Internet if required privacy challenges were not contemplated and anticipated [123, 124].

Regarding Renewable Energies, in Singapore, the commissioning of solar panels as a source was managed through a *Smart Grid* [125, 126]. There, alternative energy production means (solar panels, wind turbines, energy storage facilities, among others) were linked; power lines for the transmission of the produced energy, coupled with a controller that distributes the required electricity to the residences, balances the electricity consumption allowing seamless interconnectivity to increase efficiency and reliability of the network.

Therefore, obtaining the highest efficiency in photovoltaic systems allowed reducing the energy demand from a conventional power grid, which consequently contributed to reducing the pollution rates. Adequate orientation and tilt angle of the solar panels, on the other hand, ensured their higher efficiency and, as a result, higher exploitation of this energy to meet the *SSC* purposes, [127, 128].

In order to reduce the excess of solid waste and to establish an impulse towards green energy, an idea was found as an isolated experience that was successfully applied in a poultry farm in Singapore; an anaerobic biodigester system was implemented there to produce biogas from poultry manure, which allowed it to generate electricity, guaranteeing self-sufficiency and sustainability of the farm. The excess electricity that was not consumed was sold to the city's power grid [129, 130].

On the other hand, and related to the energy issue, water management was particularly important to meet daily demand, [131]. In this regard, Singapore inaugurated the largest seawater desalination plant in Asia to supply 25% of the population by implementing innovative technologies based on ultrafiltration membranes. Water quality monitoring is verified by a sensor network to ensure that the liquid is in the best conditions to be consumed, [132]. Moreover, the city has set out to reduce concentration rates of fine particles in the air, the level of sulfur dioxide, and the measurement of parameters such as noise, humidity, and temperature, [133].

In another approach, that of urban agriculture, hydroponic vegetable cultivation in Singapore was a measure used to supply the demand for food in the city. The use of sensors, actuators, and microprocessors to control the water quality and luminosity of these crops proved to be a self-sustainable and low-cost source that could significantly reduce the labor and operating costs usually demanded, while increasing livestock production and profitability, contributing to the sustainability and livability of the city, [134].

12.4.4 Economic Development

The key to creating innovation policies in Singapore -from 2006 to 2010 when the science and technology plan was implemented was based on establishing the appropriate means so that economic benefits could be obtained from such innovation, and the academic and scientific spheres could be generously supported to positively and gradually impact the city, [135].

In Directive Intervention, looking to achieve predetermined outcomes by making changes in investment and production patterns in selected industries, decisively promoted high-tech economy. The Singapore government provided funds for research development and innovation ($R + D + I$), established public research facilities, and helped transfer the output to the private sectors, [135, 136]. It is evident that this $R + D + I$ capability has fostered linkages between industry and science: public–private research partnerships such as the Keppel-*NUS* Corporate Laboratory with the Advanced Semiconductor Joint Laboratory Institute of Microelectronics; and the *A*STAR* Aerospace Research, Advanced Remanufacturing, and Technology Center, *LUX* photonics; or in healthcare with diabetes and gastric cancer consortia, as well as patient care logistics in hospitals, [137].

Singapore's statistics are compelling: it is considered the third richest country in the world based on its *GDP* per capita; it is the fourth-largest host of inward Foreign Direct Investment (*FDI*) flows in the world in 2018, after United States, China, and Hong Kong; it is a maritime hub connecting more than 600 ports and 120 countries; its strategic location allows it to serve as headquarters for 37,000 international companies; it has a low corporate tax rate and high level of tax benefits. However, the trade war between the United States and China, and the cyclical global slowdown in the electronics industry in 2019 due to the outbreak of COVID-19 will cause *GDP* to shrink by 3. 5% over 2020, although 3% growth is forecast in 2021 subject to global economic recovery and the size and potential of sectors such as aviation and defense, energy and environment, healthcare and *MedTech* based on *ICT* intensity and digital technologies, as well as the multiple infrastructure projects such as the new Changi airport terminal or the high-speed train between Singapore and Kuala Lumpur, [138].

12.4.5 Intellectual Capital

Singapore is a City-State whose government has sought to improve the city through strategic scientific and technological objectives—where the notion of robotics, for example, is not fundamental—[137]; four technological domains have been identified: Engineering and Advanced Manufacturing; Health Sciences and Biomedicine; Sustainability and Urban Solutions; Digital and Service Economy. The idea of integrating all the capabilities in these *ICT*-based domains is feasible because of the exponential expansion of computing power resulting in ubiquitous connectivity that modifies business models in all industries and tends to create a completely new digital economy. This, together with technological innovations in artificial intelligence and deep machine learning, can generate digital disruptive solutions that will translate into improved productivity at all levels. Given the above, Singapore has aimed to develop Strategic Research Programs, for example, Intelligent Systems: Health *ICT*, Data Science; or Emerging Areas and Industrial Alignment.

It is worth noticing, firstly, that Singapore universities have steadily risen in the world rankings since 2015, thus improving their influence and impact on international research particularly the *NUS*, and *NUT*; Table 12.3 shows this trend for 2021.

In addition, the *ICIM*—produced by the Business School of the University of Navarra—*IESE* Cities in Motion, is an aggregate indicator (101 indicators) that have made it possible to measure the sustainability of the world's main cities, as well as the quality of life of their inhabitants. Its orientation, therefore, is to lead a different type of economic and social development, which involves the creation of a global city but is based on promoting entrepreneurship, innovation, and social justice. The approach of focusing on *the Smart Governance* concept is more plausible in terms of the emergence of COVID 19, setting *"urban resilience"* as an element to elucidate the behavior against the capacity of cities to overcome traumatic circumstances and that will be part of the strategic reflections on urban agenda. that have left decimated the one corresponding to the '*Cities in Motion*' initiative, [139].

Table 12.3 World ranking of Singapore's universities

University	Status	Research output	Overall score	QS world university rankings	QS WUR ranking by subject
National University of Singapore (NUS)	Public	Very high	91.5	11	1
Nanyang Technological University. (NUT)	Public	Very high	89.9	12	1

Criteria: Academic Reputation, Citations per Faculty, Employer Reputation, Faculty Student Ratio, International Faculty Ratio, International Students Ratio

Table 12.4 Singapore's world ranking according to the ICIM

City	Population by size	Global population 2017	Global population 2018	Global population 2019
Berlin-Germany	1	4	8	7
Singapur-Singapur	2	8	6	9
Hong Kong- China	3	27	14	10
Chicago- EEUU	4	18	13	13
Washington-EEUU	5	15	20	15

Performance in nine dimensions: human capital, social cohesion, economy, governance, environment, mobility, and transportation, urban planning, international projection, and technology

The ICIM ranking for 2020, shown in Table 12.4, highlights Singapore's place among the so-called large cities (26): those with a population of around 5–10 million inhabitants.

Following the same line, the main research works of public and private institutions dedicated to the study of digitization of an economy and society, use as a reference the global indicator *Digital Economy and Society Index* (DESI), developed by the European Commission, which averages five main dimensions: Connectivity (25%); Human Capital (25%); Internet Use (15%); Digital Technology Integration (20%); and digital public services (15%). However, the results of this indicator and its monitoring are relevant as they serve to evaluate the success of the Strategy for a Digital Single Market, [140].

12.4.6 Quality of Life (Health and Sanitation, Universal Access, Deployment of Several ICT Measures for the Enjoyment of the City)

Research has been developed at all levels, but the one that stands out is the one whose objective was to find the thermal comfort of a person in his workplace or home, whose reflection is found in a daily energy use/consumption to maintain such comfort [141, 142]. The prediction, using Big Data, served four public residential buildings in the city to find an optimal balance between energy use and thermal comfort to reduce energy consumption because high consumption was observed regarding the use of ventilation systems [143–146].

Another focus of the data was related to the study of failures of the Singapore railway system so to determine the system reliability and decide the maintenance cycle based on the estimation of the meantime until the occurrence of a system failure, [147].

On the other hand, performing fast processing of acquired data from all kinds of sensors deployed in the city has required, on some occasions, to make a simulation of these data to previously know the tendency towards a specific type of result to make the most accurate decision according to the need to be solved. Based on this, Big Data has been a tool that processes data from sensors located in strategic places [148, 149]; and the use of this data has been focused on the solution and ease in the collection and treatment of waste to improve the environment quality [150]; through the creation of databases that store the places where this waste is generated, and with the implementation of ecological networks, viable commercial industrial symbioses have been established to obtain a benefit from the collected waste [151, 152].

As a consequence of what was exposed in the chapter, the evaluation of Singapore city using the adapted *GMSDIV* model, allows to observe the close collaboration of the *SSC* paradigm with the *IoT* paradigm by assimilating the technological devices (smartphones, tablets) used by citizens as the best sensor network of a city (Participatory Sensing Platform or *PS*); thanks to the fast processing they have it is possible to provide real-time information on certain events that occur daily. The multiple applications of this type of platforms support the development of applied measurement systems; for example, in transportation travel (*TQMS*), large-scale data and coverage are provided at any location. It is shown that the *PS* paradigm can be determinant in the assessment of quality of life, travel in this case [153, 154].

On the other hand, it is deduced from what has been developed in this chapter that the *IoT* paradigm requires the existence of an adequate information flow that can be used to know in real time the state of a particular urban area; but the interconnection of sensors and the data transfer they produce to the cloud are still challenges for the *SSC* [155–157]. In that case, custom sensor nodes with hybrid cloud infrastructure (*SENSg*) used in Singapore since 2015 to perform large-scale experiments on environmental factors enabled the collection of corresponding daily activity data from up to 50,000 students. The objective of such a project was to find a solution facing large-scale urban deployment of environmental sensors that, supported with the design of an integrated algorithm, reduced the amount of data to be moved and processed on central servers.

These solutions solved the problems in the urban deployment of environmental sensors, and with the design of integrated algorithms demonstrated an important advance in the ability to implement machine learning code and Deep Learning Architecture. However, to solve issue of *IoT*-side computing customers and better the quality of services, the processing must be made direct on the devices of network -Edge Computing-, and the resources-driven devices should be interconnected in an interlinked *IoT*, [158–160].

However, the ongoing discussion about the ability of public cloud computing providers to expand an *IoT* ecosystem that considers third-party access to their infrastructure would allow combining data and computing resources operating on *IoT* devices. Integration will always be an issue between architectures, but as long as cloud computing resources are fast and flexible, sensors and devices established before integrating data and their cloud offerings would allow them to be distributed across any cloud resource and reduce such inconsistencies positively impacting *SSC*—this idea helped in Japan to detect radiation and radiation maps during earthquakes—and can be extended to consumers who wish to store *IoT* packages in the cloud on the fly [161], but at no cost.

12.5 Discussion

The conceptualization and realization of a *SSC* involve technological, scientific, social and state aspects that leads to an Integrated and Complementary Network of Systems translate into Governance, Mobility, Sustainability, Economic Development, Intellectual Capital, and Quality of Life. The orientation, however, involves adding to the technological utopia a new concept of economic and social development that entails the creation of an *SSC* attached to the promotion of entrepreneurship, innovation, and social justice.

In this perspective, the chapter states that the collaboration found between the *IoT* paradigm and the *SSC* and their success stories, has emphasized the need to have an adequate flow of information to know in real time the state of a particular urban area. However, there are still challenges in the interconnection of highly sensitive sensors, as well as the transfer of the data produced to the cloud. In that vision, a hybrid cloud architecture would allow large-scale processing and prediction of environmental factors from the focal collection of daily activity data. Proportionally, the design of the general integrative algorithm—as well as the specific algorithms—should be oriented to reduce the amount of data to be moved and processed on central servers, and this leads to the systematic implementation of machine learning code and Deep Learning Architecture. Hence, future works involves keep solving the trust issue of *IoT*-side computing customers, improve the quality of services of the *IoT*'s edge computation platform using direct processing on the devices of network -without exit to internet-, and that the resources-driven devices can be interconnecting and interlinking *IoT* local system.

Furthermore, the chapter reaffirms that the ongoing discussion about the ability of public cloud computing providers to expand an *IoT* ecosystem that considers third-party access to their infrastructure should enable the combination of data and computing resources operating on *IoT* devices. Integration will always be an issue between architectures, but as long as cloud computing resources are fast and flexible, sensors and devices established before integrating data and their cloud offerings can be distributed across any cloud resource, reducing such inconsistencies and positively impacting their management in the *SSC*.

That said, the cost of technological artifacts to address *SSC* construction alternatives will always be high. But the baseline given in the chapter opens up prospects for research, development and innovation (R + D + I) to continue exploring technical-scientific alternatives focused on reliability. These, in addition to the robustness of the Internet, have to consider the infrastructure to energetically support sensors, networks, as well as the corresponding management involved in Big Data, Data Mining, among other technologies, and have to establish trends of the usefulness of the information collected, its risks and vulnerabilities. In this sense, the convergence of paradigms such as *IoT*, Sustainability and smart cities will be successful.

On the other hand, *urban resilience* as a rational element of behavior facing the capacity of cities to overcome traumatic circumstances such as the catastrophic effects of COVID-19, will be part of the strategic reflections on the sustainable and intelligent urban agenda in the future.

This document provides considerations that can be harmonized, through situated research, with the diagnostics provided by urban resilience profiling tools (*CRPT*), self-assessment tools for resilience to localized disasters (Scorecard), or city scanning tools (City Scan Tool); to better understand, plan and implement the *SSC* in the face of risks, vulnerabilities or specific local variables. The methodology and the *SSC* assessment model shown here can be aligned, then, with indicators established in standards such as ISO 37122 and 37123 (smart city and resilient city indicators respectively).

12.6 Conclusion

Singapore, as a City-State has defined to prioritize technological domains such as Engineering and Advanced Manufacturing; Health Sciences and Biomedicine; Sustainability and Urban Solutions; Digital and Service Economy. Integrating all its capabilities in these domains, based on *ICT*, viable by developments in artificial intelligence and deep machine learning, generating digital disruptive innovations that fit an *SSC*.

However, this was largely due to the development of a critical mass -converted into high-level intellectual capital- from the world-class universities it hosts. Thus, the application of a concrete and simple evaluation model can yield a quantitative assessment in terms of *GMSDIV* (Governance, Mobility, Sustainability, Economic Development, Intellectual Capital, Quality of Life), which classifies Singapore as a resilient *SSC*; but its strengthening will be guaranteed by the qualitative and committed capacity of its citizens, which implies future scopes enriching these models with new categories of analysis and evaluation.

References

1. Goldmark, P.: The new rural society. In: Paper presented at the National Cable Television Association Annual Convention, Chicago, Illinois, (17–20 May 1972)
2. Grabow, S.: Frank Lloyd Wright and the American city: the broadacres debate. J. Am. Inst. Plann. **43**(2), 115–124 (1977). https://doi.org/10.1080/01944367708977768
3. Watson, J.: The suburbanity of Frank Lloyd Wright's Boadacre City. J. Urban Hist. **45**(5), 1006–1029 (2019)
4. Fishman, R.: Urban utopias in the twentieth century. In: Howard, E., Wright, F.L., Corbusier, L. (eds.) Cambridge Massachusetts, MIT Press (1982)
5. Jacobs, J.: Muerte y vida de las grandes ciudades. Capitán Swing Libros S.l., (2011)
6. McLuhan, M.: La Galaxia Gutenberg. Génesis del "Homo Typographicus" Barcelona, Planeta- De Agostini S.A., (1985)
7. Angelidou, M.: Smart cities: a conjuncture of four forces. Cities **47**, 95–106 (2015). https://doi.org/10.1016/j.cities.2015.05.004

8. Townsend, A.: Smart Cities: Big Data, Civic Hackers, and the Quest for a New Utopia. W.W. Norton and Company, New York (2013)
9. Hollands, R.: Critical interventions into the corporate smart city. Camb. J. Reg. Econ. Soc. **8**(1), 61–77 (2015). https://doi.org/10.1093/cjres/rsu011
10. Sikora, D.: Factores de desarrollo de las ciudades inteligentes. Revista Universitaria de Geografía, Universidad Nacional del Sur Bahía Blanca, Argentina **26**(1), 135–152 (2017)
11. Sassen, S.: La Ciudad global: Nueva York, Londres, Tokio, p. 458. Buenos Aires, Eudeba (1999)
12. Montejano, J.: "El impacto de las nuevas tecnologías en la 'explosión' de la ciudad", *URBS*. Revista de Estudios Urbanos y Ciencias Sociales **3**(1), 45–67 (2013)
13. Pineda de Alcázar, M.: La Internet de las Cosas, el Big Data y los nuevos problemas de la comunicación en el Siglo XXI. Revista Mediaciones Sociales **17**, 1–24 (2018). https://doi.org/10.5209/MESO.60190
14. Cirill, F., Gómez, D., Diez, L., Maestro, I.E., Gilbert, T.B.J., Akhavan, R.: Smart city *IoT* services creation through large-*Sc*ale collaboration. IEEE Internet of Things J. **7**(6), 5267–5275 (2020). https://doi.org/10.1109/JIOT.2020.2978770
15. Vanolo, A.: Is there anybody out there? The place and role of citizens in tomorrow's smart cities. Futures **82**, 26–36 (2016)
16. González-Bustamante, R.A., Ferro-Escobar, R., Vacca-González, H.: "Smart cities in collaboration with the internet of things". Visión Electrónica, **14**(2), 185–195 (2020). https://doi.org/10.14483/22484728.16995
17. Javaid, S., Sufian, A., Pervaiz, S., Tanveer, M.: Smart Traffic Management System Using Internet of Things, pp. 393–398 (2018). https://doi.org/10.23919/ICACT.2018.8323770
18. Montori, F., Bedogni, L., Bononi, L.: A Collaborative Internet of Things architecture for smart cities and environmental monitoring. IEEE Internet of Things J. **5**(2), 592–605 (2018). https://doi.org/10.1109/JIOT.2017.2720855
19. Evans, D.: "The Internet of Things: how the next evolution of the internet is changing everything". Cisco Internet Business Solutions Group (IBSG), p. 3 (2011)
20. Rucinski, A., Garbos, R., Jeffords, J., Chowdbury, S.: "Disruptive innovation in the era of global cyber-society: with focus on smart city efforts." 9th IEEE International Conference on Intelligent Data Acquisition and Advanced Computing Systems: Technology and Applications (IDAACS), Bucharest, pp. 1102–1104 (2017). https://doi.org/10.1109/IDAACS.2017.8095256
21. Castells, M., Hall, P.: Las Tecnópolis del mundo: la formación de los complejos industriales del S. XXI. Madrid, Alianza (1994)
22. Echeverría, J.: Los Señores Del Aire: Telépolis y El Tercer Entorno. Destino, Barcelona (1999)
23. Ascher, F.: Metapolis Ou L'avenir des Villes. Editions Odile Jacob, Paris (1995)
24. Améndola, G.: La Ciudad Postmoderna: Magia y Miedo De La Metrópolis Contemporánea. Celeste, Madrid (2000)
25. Aragona, S.: La citta virtuale: trasformazioni urbane e nuove tecnologie dell'informazione. Roma, Gangemi (1993)
26. Soja, E.W.: Postmetropolis. Critical Studies of cities and regions. Oxford, Blackwell of World Affairs (2001)
27. Dematteis, G.: Suburbanización y periurbanización. Ciudades anglosajonas y ciudades latinas. In: Monclús, F.J. (ed.), La ciudad dispersa. Suburbanización y nuevas periferias, pp. 17–33. Barcelona, España, CCCB (1998)
28. Harris, P.H.: "The technopolis phenomenon - smart cities, fast systems, global networks." Behav. Sci., **38**(2) (1992)
29. Glasmeiera, A., Christopherson, S.: Thinking about smart cities. Cambridge J. Reg. Econ. Soc. **8**, 3–12 (2015). https://doi.org/10.1093/cjres/rsu034
30. Batty, M.: Intelligent cities: using information networks to gain competitive advantage. Environment and Planning B. Plan. Des. **17**(3), 47–256 (1990). https://doi.org/10.1068/b170247

31. Laterasse, J.: The intelligent city. In: Rowe, F., Veltz, P. (eds.), Telecom, companies, territories. Paris, Presses de L'ENPC (1992)
32. Moyser, R.: Defining and Benchmarking SMART Cities. [On line]. http://www.burohappold.com/blog/arTICle/defining-andbenchmarking-smart-cities-1771/ (2013)
33. Shelton, T., Zook, M., Wiig, A.: Thinking about smart cities. Cambridge J. Reg. Econ. Soc. **8**, 13–25 (2015). https://doi.org/10.1093/cjres/rsu026
34. Mitchell, W.J.: "Ciudades inteligentes". UOcPapers: Revista sobre la Sociedad del Conocimiento, Universitat Oberta de Catalunya, **5**, 1–12 (2007)
35. Zona, A.T., Fajardo, C.H., Aguilar, C.M.: Propuesta De Un Marco General Para El Despliegue De Ciudades Inteligentes Apoyado En El Desarrollo De *IoT* En Colombia. Revista Ibérica de Sistemas e Tecnologias de Informaçión **28**(4), 894–907 (2020)
36. Batty, M., Axhausen, K.W., Giannotti, F., Pozdnoukhov, A., Bazzani, A., Wachowicz, M., Ouzounis, G., Portugali, Y.: Smart cities of the future. Euro. Phys. J. **214**, 481–518 (2012)
37. Parysek, J., Mierzejewska, L.: Spatial structure of a city and the mobility of its residents: functional and planning aspects. Bull. Geogr. Socioecon. Series **34**(34), 91–102 (2016)
38. Soyata, T., Habibzadeh, H., Ekenna, C., Nussbaum, B., Lozano, J.: Smart city in crisis: technology and policy concerns. Sustain. Cities Soc. **50**(April) (2019). https://doi.org/10.1016/j.SCs.2019.101566
39. Roller, L.H., Waverman, L.: Telecomunication infrastructure and economic development: a simultaneous approach. Am. Econ. Rev. **91**(4), 909–923 (2001)
40. Florida, R.: The rise of the creative class. Basic Books, Fondation Le Corbusier (2014)
41. Mejía, J., Quintero, S.: "Comprensión de las ciudades contemporáneas como estructuras organizacionales. Aporte a la categoría de ciudades inteligentes". Cuadernos de Vivienda y Urbanismo, **13**, 1–9 (2020). https://doi.org/10.11144/Javeriana.cvu13.ccce
42. Alderete, M.V.: ¿Qué factores influyen en la construcción de ciudades inteligentes? Un modelo multinivel con datos a nivel ciudades y países. CTS **14**(41), 71–89 (2020)
43. Anttiroiko, A.V., Valkama, P., Bailey, S.J.: Smart cities in the new service economy: building platforms for smart services. AI Soc. **29**, 323–334 (2014)
44. Beatley, T., Newman, P.: Green Urbanism Down Under: Learning from Sustainable Communities in Australia. Island Press, Washington, DC (2008)
45. Sánchez, R., Nuñez, A., Sesma, J., Bilbao, A., Mulero, R., Zulaika, U., Azkune, G., Almeida, A.: Smart cities survey: technologies, application domains and challenges for the cities of the future. Int. J. Distrib. Sens. Netw. **15**(6), 1–36 (2019). https://doi.org/10.1177/1550147719853984
46. Hodgkinson, S.: Is your city smart enough? digitally enabled cities and societies will enhance economic, social, and environmental sustainability in the urban century. OVUM report (2011)
47. Townsend, A., Maguire, R., Liebhold, M., Crawford, M.: The future of cities, information, and inclusion: a planet of civic laboratories. Institute for the Future (2010)
48. Economic Commission for Europe ECE. Summary of activities on smart sustainable cities (*SSC*) of ECE Committee on Housing and Land Management. Geneva, (2–4, Oct 2019)
49. Sosa, E., et al.: "Internet del Futuro y Ciudades Inteligentes", XV Workshop de investigadores en Ciencias de la Computación, pp. 21–28 (2013)
50. Shen, C., Zhang, K., Long, K.: "Research on hainan trusted digital infrastructure construction framework." 2020 29th Wireless and Optical Communications Conference (WOCC), Newark, NJ, USA, pp. 1–5 (2020). https://doi.org/10.1109/WOCC48579.2020.9114945
51. Patra, M.: "An architecture model for smart city using Cognitive Internet of Things(*CIoT*)", Second International Conference on Electrical, Computer and Communication Technologies (ICECCT), 1–6, (2017). https://doi.org/10.1109/ICECCT.2017.8117893
52. Long, K., Nsalo, D.F., Zhang, K., Tian, C., Shen, C.: " A CSI-based indoor positioning system using single UWB ranging correction." Sensors (Basel), **21**(19) (2021). https://doi.org/10.3390/s21196447
53. Herrera, F., et al.: "Informe sobre La tendencia inteligente de las ciudades en España", Grupo de Smart Cities/Smart Regions Colegio Oficial de Ingenieros de Telecomunicación, pp. 1–50 (2018)

54. Alkhabbas, F., Spalazzese, R., Davidsson, P.: "Architecting Emergent configurations in the Internet of Things." 2017 IEEE International Conference on Software Architecture (ICSA), Gothenburg, pp. 221–224, (2017). https://doi.org/10.1109/ICSA.2017.37
55. Merry, H.: "Population increase and the smart city,", Business Operations Blog (2017)
56. Hribar, J., DaSilva, L.: "Utilising correlated information to improve the sustainability of Internet of Things devices." 2019 IEEE 5th World Forum on Internet of Things (WF-IoT), Limerick, Ireland, pp. 805–808 (2019). https://doi.org/10.1109/WF-IoT.2019.8767256
57. Igder, S., Bhattacharya, S., Elmirghani, J.: Energy efficient fog servers for Internet of Things Information Piece Delivery (IoTIPD) in a smart city vehicular environment. International Conference on Next Generation Mobile Applications, Services, and Technologies, pp. 99–104 (2016). https://doi.org/10.1109/NGMAST.2016.17
58. Yaqoob, S., Ullah, A., Akbar, M., Imran, M., Guizani, M.: "Fog-assisted congestion avoidance *SC*heme for internet of vehicles." 2018 14th International Wireless Communications and Mobile Computing Conference (IWCMC), Limassol, pp. 618–622 (2018). https://doi.org/10.1109/IWCMC.2018.8450402
59. Bogatinoska, D., Malekian, R., Trengoska, J., Nyakom W.: Advanced sensing and internet of things in smart cities. 2016 39th International Convention on Information and Communication Technology, Electronics and Microelectronics, MIPRO, Proceedings, pp. 632–637 (2016). https://doi.org/10.1109/MIPRO.2016.7522218
60. Rizwan, P., Suresh, K., Babu, M.: Real-time smart traffic management system for smart cities by using Internet of Things and big data. International Conference on Emerging Technological Trends (ICETT), pp. 1–7 (2016). https://doi.org/10.1109/ICETT.2016.7873660
61. Enerlis, Ferrovial and Madrid Network, Ernst and Young: "Libro Blanco Smart Cities" (1st ed.), España (2012)
62. Trabuchi, S.: "Liderar la normalización del Internet de las Cosas (IoT) para construir ciudades inteligentes y sostenibles". *Cantones Sostenibles para la Costa Rica del siglo XXI*, ITU-MICITT, San José, Costa Rica (2016)
63. Lanfor, O.G., Pérez, J.F.: Implementación de un sistema de seguridad independiente y automatización de una residencia por medio del internet de las cosas, *Student Conference (CONESCAPAN)*, IEEE Central America and Panama, pp. 1–5 (2017)
64. Mora, O., Rivera, R., Larios, V., Beltrán-Ramírez, J., Maciel, R., Ochoa, A.: "A use case in cybersecurity based in Blockchain to deal with the security and privacy of citizens and Smart Cities Cyberinfrastructures." *IEEE International Smart Cities Conference (ISC2)*, Kansas City, pp. 1–4 (2018). [On line]. https://doi.org/10.1109/ISC2.2018.8656694
65. Olaya, A.: "Convocatoria VIVE DIGITAL regional". Ministerio de Tecnologías de la Información y las Comunicaciones y el Departamento Nacional de Ciencia Tecnología e Innovación, Colombia (2015)
66. Najar-Pacheco, J., Bohada-Jaime, J., Rojas-Moreno, W.: "Vulnerabilidades en el internet de las cosas". Visión Electrónica **13**(2), 312–321 (2019). https://doi.org/10.14483/22484728.15163
67. Ahlgren, B., Hidell, M., Ngai, E.: Internet of Things for smart cities: interoperability and open data. IEEE Internet Comput. **20**(6), 52–56 (2016). https://doi.org/10.1109/MIC.2016.124
68. An, J., et al.: Toward global *IoT*-enabled smart cities interworking using adaptive Seman*TIC* adapter. IEEE Internet of Things J. **6**(3), 5753–5765 (2019). https://doi.org/10.1109/JIOT.2019.2905275
69. Barrera, M., Serrato, N., Rojas, E., Mancilla, G.: Estado del arte en redes definidas por software (SDN). Visión Electrónica **13**(1), 178–194 (2019). https://doi.org/10.14483/22484728.14424
70. López-López, E.A., Álvarez-Aros, E.L.: "Estrategia en ciudades inteligentes e inclusión social del adulto mayor". Paakat: Revista de Tecnología y Sociedad, **11**(20), 1–29 (2021). https://doi.org/10.32870/Pk.a11n20.543
71. Ghosh, S.: "Smart homes: architectural and engineering design imperatives for smart city building codes". *Technologies for Smart-City Energy Security and Power (ICSESP)*, Bhubaneswar, pp. 1–4 (2018). https://doi.org/10.1109/ICSESP.2018.8376676N

72. Villanueva-Rosales, Garnica-Chavira, L., Larios, V.M., Gómez, L., Aceves, E.: "Semantic-enhanced living labs for better interoperability of smart cities solutions." *IEEE International Smart Cities Conference (ISC2),* Trento, pp. 1–2 (2016). https://doi.org/10.1109/ISC2.2016.7580775
73. Okai, E., Feng, X., Sant, P.: "Smart Cities Survey." 2018 IEEE 20th International Conference on High Performance Computing and Communications; IEEE 16th International Conference on Smart City; IEEE 4th International Conference on Data Science and Systems (HPCC/SmartCity/DSS), pp. 1726–1730 (2018). https://doi.org/10.1109/HPCC/SmartCity/DSS.2018.00282.
74. Ejaz, W., Naeem, M., Shahid, A., Anpalagan, A., Jo, M.: "Efficient energy management for Internet of Things in smart cities", IEEE Commun. Mag., pp. 84–91 (2017). https://doi.org/10.1109/MCOM.2017.1600218CM
75. Ramirez, J., Pedraza, C.: "Performance analysis of communication protocols for Internet of things platforms." 2017 IEEE Colombian Conference on Communications and Computing (COLCOM), pp. 1–7 (2017). https://doi.org/10.1109/ColComCon.2017.8088198
76. Rose, K., Eldridge S., Chapin, L.: "The internet of things: an overview. Understanding the Issues and Challenges of a More Connected World (2015)
77. López, R.A.: "Ciudad inteligente y sostenible: hacia un modelo de innovación inclusiva", 2007–3607 (2007). https://doi.org/10.32870/Pk.a7n13.299
78. Arias, G.: "Marco regulatorio para la provisión de Contenidos y Aplicaciones (PCA) y condiciones normativas para la adopción del Internet de las Cosas (IoT)". Comisión de Regulación de Comunicaciones (CRC), Colombia (2015)
79. Anuar, R.N., Osman, M.H.M., Ismail, K.: "Innovation centers in Malaysia: a proposed model." 2012 International Conference on Innovation Management and Technology Research, pp. 337–341 (2012). https://doi.org/10.1109/ICIMTR.2012.6236414
80. Estavillo-Flores, M.E.: "Internet de las cosas: retos para su desarrollo", IFT, CDMX (2016)
81. Silva, B.N., et al.: Urban planning and smart city decision management empowered by real-time data processing using big data analytics. Sensors (Switzerland) **18**(9), 6–12 (2018)
82. Millahual, C.: Arduino de cero a Experto. Buenos Aires, Argentina Six ediciones, ISBN 978–987–46518–7–7, p. 29 (2017)
83. Monteiro, A., de Oliveira, M., de Oliveira, R., da Silva, T.: Embedded application of convolutional neural networks on Raspberry Pi for SHM. Electron. Lett. **54**(11), 680–682 (2018). https://doi.org/10.1049/el.2018.0877
84. Rosli, R.S., Habaebi, M.H., Islam, M.R.: "Characteristic analysis of received signal strength indicator from ESP8266 WiFi transceiver module," *7th International Conference on Computer and Communication Engineering (ICCCE),* Kuala Lumpur, pp. 504–507 (2018). https://doi.org/10.1109/ICCCE.2018.8539338
85. Gayathri, K.: "Implementation of environment parameters monitoring in a manufacturing industry using IoT." 2019 5th International Conference on Advanced Computing and Communication Systems (ICACCS), pp. 858–862. Coimbatore, India (2019). https://doi.org/10.1109/ICACCS.2019.8728365
86. Simić, M., Stojanović, G.M., Manjakkal, L., Zaraska, K.: "Multi-sensor system for remote environmental (air and water) quality monitoring." 24th Telecommunications Forum(TELFOR), Belgrade, pp. 1–4 (2016). https://doi.org/10.1109/TELFOR.2016.7818711
87. Meyer, D., Kunz, P., Cox, K.: Waveguide-Coupled Rydberg Spectrum Analyzer from 0 to 20 GHz. Phys. Rev. Applied **15**, 014053 (2021). https://doi.org/10.1103/PhysRevApplied.15.014053
88. IMD-SUTD. Smart City Index 2020. A tool for action, an instrument for better lives for all citizen, (2020)
89. Orgaz, C.J.: Cómo es Songdo, la ciudad inteligente creada desde cero en Corea del Sur. BBC News Mundo (2021)
90. Lee, S.K., Kwon, H.R., Cho, H.A., Kim, J., Lee, D.: International case studies of smart cities: Songdo, Republic of Korea. Inter-American Developmen Bank IDB (2016)

91. Su, L., Janajreh, I.: "Wind energy assessment: Masdar city case study," 2012 8th International Symposium on Mechatronics and its Applications, pp. 1–6 (2012). https://doi.org/10.1109/ISMA.2012.6215162
92. Sankaran, V., A.: "Creating global sustainable smart cities (A Case Study of Masdar City)". First Int. Conf. Adv. Phys. Sci. Mater. J. Phys.: Conf. Series **1706**, 012141 (2020). https://doi.org/10.1088/1742-6596/1706/1/012141
93. Cugurullo, F.: Urban eco-modernisation and the policy context of new eco-city projects: Where Masdar City fails and why. Urban Stud. **53**(11), 2417–2433 (2015). https://doi.org/10.1177/0042098015588727
94. Moreno, C.: "Desarrollo De Un Modelo De Evaluación De Ciudades Basado En El Concepto De Ciudad Inteligente (Smart City)," (Tesis Doctoral), Universidad Politécnica de Madrid, Escuela técnica superior de ingenieros de caminos canales y puertos, p. 413 (2015)
95. Research Innovation Enterprise 2020, RIE2020 Plan. Winning the Future through Science and Technology
96. Índice IESE, Cities in Motion 2020. https://doi.org/10.15581/018.ST-542
97. Málaga innovadora: Propuesta para medir la Ciudad Inteligente. Editora: Fundación CIEDES. Málaga, 17 cuadernos II Plan Estratégico de Málaga (2018)
98. Agencia Vasca de Internacionalización, Basque Trade and Investment S.A. Informe País Singapur, pp. 1–33, junio 2020
99. Houghton, C., Reiners, J., Lim, J.: "Transporte inteligente," Transp. Intel. Cómo Mejorar la Movilidad en las Ciudades, p. 24 (2009)
100. Gutiérrez B, J.: "International Case Studies of Smart Cities: Santander, Spain," Washington, D.C., (2016)
101. Seng, C.E.: "Singapore's smart nation program—Enablers and challenges," *11th Syst. Syst. Eng. Conf.*, pp. 1–5 (2016)
102. Harrison, C., Eckman, B., Hamilton, R., Hartswick, P.: "Foundations for smarter cities," **54**(4), 1–16 (2010)
103. Fundación TELEFÓNICA, "Smart Cities: un primer paso hacia la internet de las cosas," Editor. Ariel, pp. 13–16 (2011)
104. Kamargianni, M., Li, W., Matyas, M., Schäfer, A.: A critical review of new mobility services for urban transport. Transp. Res. Procedia **14**, 3294–3303 (2016)
105. Li, G., Yu, L., Ng, W.S., Wu, W., Goh, S.T.: "Predicting home and work locations using public transport smart card data by spectral analysis." IEEE Conf. Intell. Transp. Syst. Proc. ITSC **2015**(Octob), 2788–2793 (2015)
106. Othman, N.B., Legara, E.F., Selvam, V., Monterola, C.: Simulating congestion dynamics of train rapid transit using smart card data. Procedia Comput. Sci. **29**, 1610–1620 (2014)
107. Sebhatu, S.P., Enquist, B.: Sustainable public transit service value network for building living cities in emerging economies: multiple case studies from public transit services. Procedia—Soc. Behav. Sci. **224**, 263–268 (2016)
108. Di Pasquale, G., Dos Santos, A.S., Leal, A.G., Tozzi, M.: Innovative public transport in Europe, Asia and Latin America: a survey of recent implementations. Transp. Res. Procedia **14**, 3284–3293 (2016)
109. Regional metropolitano de Santiago, G.: "Revisión y Actualización del Plan Maestro de Ciclovías y Plan de Obras" (2012)
110. Goletz, M., Feige, I., Heinrichs, D.: "What Drives mobility trends: results from case studies in Paris, Santiago de Chile, Singapore and Vienna," Transp. Res. Procedia, **13**(Pucher 2010), 49–60 (2016)
111. Xiao, Y. et al.: "Transportation activity analysis using smartphones," IEEE Consumer Commun. Netw., pp. 60–61 (2012)
112. Nazir, S., Wong, Y.S.: Energy and pollutant damage costs of operating electric, hybrid, and conventional vehicles in Singapore. Energy Procedia **14**, 1099–1104 (2012)
113. Nian, V., Hari, M.P., Yuan, J.: The prospects of electric vehicles in cities without policy support. Energy Procedia **143**, 33–38 (2017)

114. Chia, M.Y.W., Krishnan, S., Zhou, J.: "Challenges and opportunities in infrastructure support for electric vehicles and smart grid in a dense urban Environment-Singapore," IEEE Int. Electr. Veh. Conf. IEVC 2012 (2012)
115. Lokesh, B.T., Tay, J., Min, H.: A Framework for Electric Vehicle (EV) charging in Singapore The 15th International Symposium on district heating and cooling assessing. Energy Procedia **143**, 15–20 (2017)
116. Kumar, K.N., Tseng, K.J.: Impact of demand response management on chargeability of electric vehicles. Energy **111**, 190–196 (2016)
117. Leurent, F.: Modeling transportation systems involving autonomous vehicles: a state of the art. Transp. Res. Procedia **27**, 215–221 (2017)
118. Krishnasamy, C., Unsworth, C., Howie, L.: The patterns of activity, and transport to activities among older adults in Singapore. Hong Kong J. Occup. Ther. **21**(2), 80–87 (2011)
119. Khansari, N., Mostashari, A., Mansouri, M.: Conceptual modeling of the impact of smart cities on household energy consumption. Procedia Comput. Sci. **28**(Cser), 81–86 (2014)
120. Alvina, P., Bai, X., Chang, Y., Liang, D., Lee, K.: Smart community based solution for energy management: an experimental setup for encouraging residential and commercial consumers participation in demand response program. Energy Procedia **143**, 635–640 (2017)
121. Chuan, L., Ukil, A., Member, S.: Modeling and validation of electrical load pro fi ling in residential buildings in Singapore. IEEE Trans. Power Syst. **30**(5), 1–10 (2014)
122. Wouters, C.: Towards a regulatory framework for microgrids—the Singapore experience. Sustain. Cities Soc. **15**, 22–32 (2015)
123. Kumar, A., Kar, P., Warrier, R., Kajale, A., Panda, S.K.: Implementation of smart LED lighting and efficient data management system for buildings. Energy Procedia **143**, 173–178 (2017)
124. Fernandes, R.F., Fonseca, C.C., Brandão, D., Carlos, S.: "Flexible Wireless Sensor Network for smart lighting applications." (2014)
125. Kazmi, H., Mehmood, F., Amayri, M.: "Smart home futures: algorithmic challenges and opportunities," *14th International Symposium on Pervasive Systems, Algorithms and Networks and 11th International Conference on Frontier of Computer Science and Technology and 7 Third International Symposium of Creative Computing (ISPAN-FCST-ISCC)*, pp. 441–448 (2017)
126. Bhati, A., Hansen, M., Chan, C.M.: Energy conservation through smart homes in a smart city: a lesson for Singapore households. Energy Policy **104**(February), 230–239 (2017)
127. Bugeja, J., Jacobsson, A., Davidsson, P.: "On privacy and security challenges in smart connected homes," Proceedings—*2016 European Intelligence and Security Informatics Conference, EISIC 2016*, pp. 172–175 (2017)
128. Wongvises, C., Khurat, A., Fall, D., Kashihara, S.: "Fault tree analysis-based risk quantification of smart homes," *2nd International Conference on Information Technology (INCIT)*, pp. 1–6 (2017)
129. Brenda, L.G.H.: "Grid Code Compliance for Grid-Connecting a PV System to an Existing Facility in Singapore," pp. 530–533 (2016)
130. Jiang, B.T., et al.: "A Microgrid Test Bed in Singapore", pp. 74–82 (2017)
131. Khoo, Y.S., Reindl, T., Aberle, A.G., Member, S.: Optimal orientation and tilt angle for maximizing in-plane solar irradiation for PV applications in Singapore. IEEE J. Photovoltaics **4**(2), 647–653 (2014)
132. Zi'An, W., King Ho, H.L.: "Household solar 'plant' for Singapore," 2016 Asian Conference on Energy, Power and Transportation Electrification (ACEPT), pp. 1–4 (2016). https://doi.org/10.1109/ACEPT.2016.7811504
133. Yadav, S.D., Kumar, B., Thipse, S.S.: "Biogas purification: producing natural gas quality fuel from biomass for automotive applications," *International Conference on Energy Efficient Technologies for Sustainability, ICEETS*, pp. 450–452 (2013)
134. Heng, D.L.K.: Bio gas plant green energy from poultry wastes in Singapore. Energy Procedia **143**, 436–441 (2017)
135. Xi, X., Leng, K.: Using system dynamics for sustainable water resources management in Singapore. Procedia Comput. Sci. **16**, 157–166 (2013)

136. Wang, J.: Innovation and government intervention: a comparison of singapore and Hong Kong. Res. Policy **47**(2), 399–412 (2018)
137. Lee, S.M., Trimi, S.: Innovation for creating a smart future. Suma Negocios **3**(1), 1–8 (2018)
138. Tan, E., So, H.: Rethinking the impact of activity design on a mobile learning trail: the missing dimension of the physical affordances. IEEE Trans. Learn. Technol. **8**(1), 98–110 (2015)
139. Yang, W., Wong, N.H., Lin, Y.: Thermal comfort in high-rise urban environments in Singapore. Procedia Eng. **121**, 2125–2131 (2015)
140. Deb, C., Eang, L., Yang, J., Santamouris, M.: Forecasting energy consumption of institutional buildings in Singapore. Procedia Eng. **121**, 1734–1740 (2015)
141. Chaudhuri, T., Soh, Y.C., Li, H., Xie, L.: "Machine Learning Based Prediction of Thermal Comfort in Buildings of Equatorial Singapore," pp. 72–77 (2017)
142. Chiu, P.H., et al.: CFD methodology development for Singapore green mark building application. Procedia Eng. **180**, 1596–1602 (2017)
143. Wong, N.H., Tan, E., Gabriela, O., Jusuf, S.K.: Indoor thermal comfort assessment of industrial buildings in Singapore. Procedia Eng. **169**, 158–165 (2016)
144. Happle, G., Wilhelm, E., Fonseca, J.A., Schlueter, A.: Determining air-conditioning usage patterns in Singapore from distributed, portable sensors. Energy Procedia **122**, 313–318 (2017)
145. Liu, Y., Wu, Y.: "Smart maintenance via dynamic fault tree analysis: a case study on Singapore MRT system," *47th Annual IEEE/IFIP International Conference on Dependable Systems and Networks (DSN)*, (2017)
146. Kong, X. et al.: "Mobility dataset generation for vehicular social networks based on floating car Data." IEEE Trans. Vehicular Technol. (2017)
147. Al Nuaimi, E., Al Neyadi, H., Mohamed, N., Al-jaroodi, J.: "Applications of big data to smart cities," J. Internet Serv. Appl., (2015)
148. Jara, A.J., Genoud, D., Bocchi, Y.: "Big Data in smart cities: from poisson to human dynamics," *28th International Conference on Advanced Information Networking and Applications Workshops* (2014)
149. Laohalidanond, K., Chaiyawong, P., Kerdsuwan, S.: Municipal solid waste characteristics and green and clean energy recovery in Asian Megacities, vol. 79. Elsevier B.V. (2015). https://doi.org/10.1016/j.egypro.2015.11.508
150. Bin, S., Zhiquan, Y., Sze, L., Jonathan, C., Koh, D., Kurle, D.: A Big-Data analytics approach to develop industrial symbioses in large cities. Procedia CIRP **29**, 450–455 (2015)
151. Song, B., Yeo, Z., Kohls, P., Herrmann, C.: Industrial symbiosis: exploring Big-Data approach for waste stream discovery. Procedia CIRP **61**, 353–358 (2017)
152. Estrin, D.: Participatory sensing: applications and architecture [Internet Predictions]. IEEE Internet Comput. **14**(1), 12–42 (2010)
153. Xiao, Z., Lim, H., Ponnambalam, L.: Participatory sensing for smart cities: a case study on transport trip quality measurement. IEEE Trans. Industr. Inf. **13**(2), 759–770 (2017)
154. Ng, S.T., Xu, F.J., Yang, Y., Lu, M.: A Master data management solution to unlock the value of big infrastructure data for smart, sustainable and resilient city planning. Procedia Eng. **196**(June), 939–947 (2017)
155. Wan, J., Li, D., Zou, C., Zhou, K.: "M2M communications for smart city: An event-based architecture," *Proc.—2012 IEEE 12th Int. Conf. Comput. Inf. Technol. CIT*, pp. 895–900 (2012)
156. Schmitt, A.: Dynamic bridge generation for IoT data exchange via the MQTT protocol. Procedia Comput. Sci. **130**, 90–97 (2018). https://doi.org/10.1016/j.procs.2018.04.016
157. Wilhelm, E., et al.: Wearable environmental sensors and infrastructure for mobile large-scale urban deployment. IEEE Sens. J. **16**(22), 8111–8123 (2016)
158. Kyaw, T.Y., Ng, A.K.: Smart aquaponics system for urban farming. Energy Procedia **143**, 342–347 (2017)
159. Alazab, M., Manogaran, G., Montenegro-Marin, C.E.: Trust management for internet of things using cloud computing and security in smart cities. Cluster Comput. (2021). https://doi.org/10.1007/s10586-021-03427-9

160. Xin, Q., Alazab, M., García, V., Montenegro-Marin, C.E., González-Crespo, R.: "A deep learning architecture for power management in smart cities," to appear in *Springer Nature* (2021)
161. Alam, T.: Cloud-based IoT applications and their roles in smart cities. Smart Cities **4**, 1196–1219 (2021). https://doi.org/10.3390/smartcities4030064

Printed in the United States
by Baker & Taylor Publisher Services